New Trends in Polymer Chemistry and Characterization

MATERIALS RESEARCH SOCIETY
SYMPOSIUM PROCEEDINGS VOLUME 1613

New Trends in Polymer Chemistry and Characterization

Symposium held August 11–15, 2013, Cancún, México

EDITORS

Lioudmila Fomina

Instituto de Investigaciones en Materiales
Universidad Nacional Autónoma de México
México City, México

Gerardo Cedillo Valverde

Instituto de Investigaciones en Materiales
Universidad Nacional Autónoma de México
México City, México

María del Pilar Carreón Castro

Instituto de Ciencias Nucleares
Universidad Nacional Autónoma de México
México City, México

Materials Research Society
Warrendale, Pennsylvania

CAMBRIDGE
UNIVERSITY PRESS

CAMBRIDGE UNIVERSITY PRESS
Cambridge, New York, Melbourne, Madrid, Cape Town,
Singapore, São Paulo, Delhi, Mexico City

Cambridge University Press
32 Avenue of the Americas, New York, NY 10013-2473, USA

www.cambridge.org
Information on this title: www.cambridge.org/9781605115900

Materials Research Society
506 Keystone Drive, Warrendale, PA 15086
http://www.mrs.org

First published 2014

CODEN: MRSPDH

ISBN: 978-1-60511-590-0 Hardback

CONTENTS

BIOPOLYMERS

POLYMER CHARACTERIZATION

RHEOLOGY IN POLYMERS

PREFACE

This volume contains papers presented at Symposium 4D, "New Trends in Polymer Chemistry and Characterization" of the XXII International Materials Research Congress organized by the Sociedad Mexicana de Materiales A. C. in collaboration with the Materials Research Society (MRS), USA, which was held from August 11 - 15, 2013, in Cancún, Mexico. The symposium was devoted to fundamental and technological applications of polymeric materials, and continued the tradition of providing a forum for scientists from various backgrounds with a common interest in the development and use of polymeric materials to come together and share their findings and expertise.

The papers contained in this volume are a collection of invited and contributed papers. This year, the symposium was attended by participants from Brazil, Canada, Chile, China, Colombia, Czech Republic, France, Germany, Japan, Mexico, Puerto Rico, Saudi Arabia and the United States. All papers have been thoroughly reviewed by at least two referees and edited to the standards of the Materials Research Society. We are grateful to all the authors who made additional efforts to prepare their manuscripts.

The "New Trends in Polymer Chemistry and Characterization" symposium has been held for the last 10 years with the objective of presenting overviews and recent investigations related to polymer engineering, polymer physics, and polymer chemistry both in academia and industry. This symposium provides a forum to debate with some of the most distinguished scientists and engineers in the international polymer science community. The topics included step-growth and chain-growth polymerizations, macromolecular architecture and topology, well-defined and uniform polymers, building blocks for nanoscience, polymer characterization by new and combined techniques, polymer morphology, dendrimers and hyperbranched polymers, smart and functional polymers, composite materials and coatings, computer-aided design and modeling of polymers and polymer forming reactions, simulation methods in polymer chemistry and physics, and environment friendly polymers and biopolymers.

The organizing committee gratefully acknowledges the enthusiastic cooperation of all symposium participants. The financial support of the Instituto de Investigaciones en Materiales, Universidad Nacional Autónoma de México (IIM-UNAM, México) is also acknowledged. We hope that all readers will come to consider the "New Trends in Polymer Chemistry and Characterization" symposium in Cancún, Mexico, as a suitable forum to present the results of their recent research and experience.

Lioudmila Fomina
Gerardo Cedillo Valverde
María del Pilar Carreón Castro

January 2014

MATERIALS RESEARCH SOCIETY SYMPOSIUM PROCEEDINGS

MATERIALS RESEARCH SOCIETY SYMPOSIUM PROCEEDINGS

Volume 1664E — Elastic Strain Engineering for Unprecedented Materials Properties, 2014, J. Li, E. Ma, Z. W. Shan, O.L. Warren, ISBN 978-1-60511-641-9

Prior Materials Research Symposium Proceedings available by contacting Materials Research Society

Polymer Synthesis

Mater. Res. Soc. Symp. Proc. Vol. 1613 © 2014 Materials Research Society
DOI: 10.1557/opl.2014.152

Rhodium-Catalyzed Oxidative Polycoupling of Phenylpyrazole and Internal Diynes: A New Polymerization Route for Atom-Economical Synthesis of Poly(pyrazolylnaphthalene)s

Yajing Liu, Meng Gao, Jie Li, and Ben Zhong Tang
Department of Chemistry, Institute for Advanced Study and Division of Biomedical
Engineering, The Hong Kong University of Science & Technology, Clear Water Bay, Kowloon,
Hong Kong, China. Email: tangbenz@ust.hk

ABSTRACT

A new route for atom-economical synthesis of functional polymers was developed. Oxidative polycoupling of 3,5-dimethyl-1-phenylpyrazole with 4,4'-(α,ω-alkylenedioxy) bis(diphenylacetylene)s and 1,2-diphenyl-1,2-bis[4-(phenylethynyl)phenyl]ethene, respectively, were catalyzed by [Cp*RhCl$_2$]$_2$, 1,2,3,4-tetraphenylcyclopenta-1,3-diene and copper(II) acetate in dimethylformamide under stoichiometric imbalance conditions, affording soluble poly(pyrazolylnaphthalene)s in satisfactory yields (isolation yield up to 82%) with high molecular weights (M_w up to 35700). All the polymers were thermally stable, losing little of their weight at high temperatures of 323–422 °C. They possessed good film-forming property and their thin solid films showed high refractive indices (RI = 1.747–1.593) in a wide wavelength region of 400–1000 nm. The polymer carrying tetraphenylethene unit displayed a phenomenon of aggregation-induced emission and showed enhanced light emission in the aggregated state.

INTRODUCTION

Development of new methodology for the synthesis of functional polymers is an important research area in macromolecular science. Olefins have been the main sources of monomers, whose addition polymerization yield polymers with electronically saturated backbones that are commonly used as commodity materials. Polymerization of acetylenic monomers can generate polymers with π-conjugated backbones that are expected to be electronically active. Indeed, polyacetylene was found to exhibit metallic conductivity upon doping by Shirakawa, MacDiarmid and Heeger in the 1970s.[1] This seminal discovery has triggered great efforts in utilizing alkynes as building blocks to construct functional polymers. As a result, a large number of π-conjugated polymers has been synthesized from alkyne monomers.[2] Among various employed methods, the polycyclotrimerization or Diels- Alder reaction could afford polymers with stable benzene rings (Scheme 1),[3] and hence broad applications in organic light-emitting diodes,[4] photovoltaic cells[5] and sensors.[6]

Highly substituted naphthalene derivatives are well-known for their high thermal stability and unique electro- and photochemical properties, as well as their potential uses as organic semiconductors and luminescent materials.[7] Recently, we found that the rhodium-catalyzed oxidative polycoupling of arylboronic acids and internal diynes proceeded smoothly under stoichiometric imbalance-promoted conditions,[8] affording poly(naphthalene)s with moderate molecular weights in satisfactory yields (Scheme 1).[9] Thanks to their high aromatic content, the resulting polymers showed high thermal stability and were light refractive. Although the polymerization reactions generated boronic acids as byproducts, we would like to develop more "green" methodologies for atom-economical synthesis of functional polymers with high molecular weights for useful practical applications.

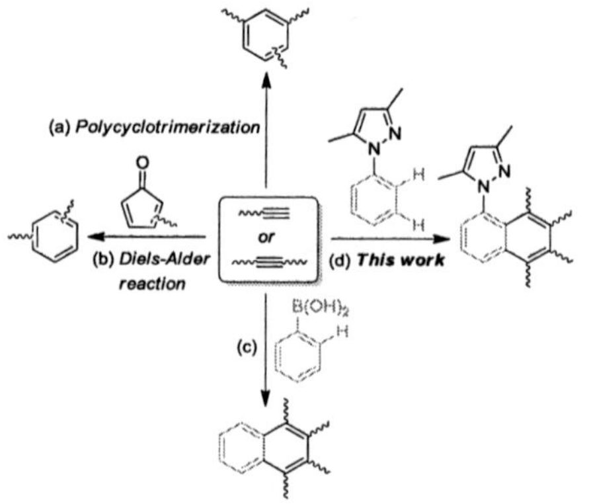

Scheme 1 Synthetic routes to polyarylenes by (a) alkyne polycyclotrimerization, (b) Diels-Alder reaction of alkyne with cyclopenta-2,4-dienone and oxidative polycouplings of (c) phenylboronic acid with internal alkyne and (d) phenylpyrazole with internal alkyne.

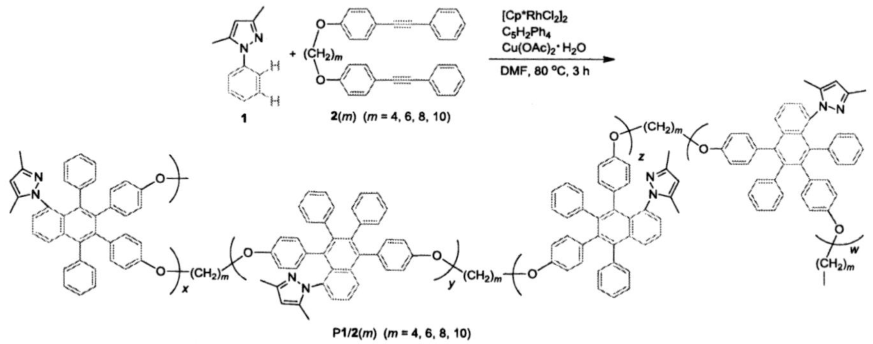

Scheme 2 Rhodium-catalyzed oxidative polycoupling of phenylpyrazole **1** and internal diynes **2**(*m*).

Satoh and Miura recently reported a promising method for the preparation of pyrazolylnaphthalenes by rhodium-catalyzed oxidative coupling of phenylpyrazoles and internal alkynes.[10] This reaction is very interesting as the regioselective C–H bond activation or cleavage can be realized by the proximate effect of pyrazole ring. Thus, this functional group will not be eliminated during the reaction. In other words, such reaction is efficient in atom economy and the products are high in purity. More importantly, the presence of excess amount of phenylpyrazole in the reaction mixture can improve the reaction efficiency significantly, suggesting that it is promising to be developed into an atom-economical polymerization tool that operates under stoichiometric imbalance conditions. However, such possibility has rarely been explored though a large number of low molecular weight pyrazolylnaphthalenes has been prepared via such method. In this paper, we took such challenge and showed that the

4

poly(pyrazolylnaphthalene)s P1/2(*m*) and P1/3 (Scheme 2 and 3) obtained under optimized conditions exhibited novel thermal, photonic and photophysical properties.

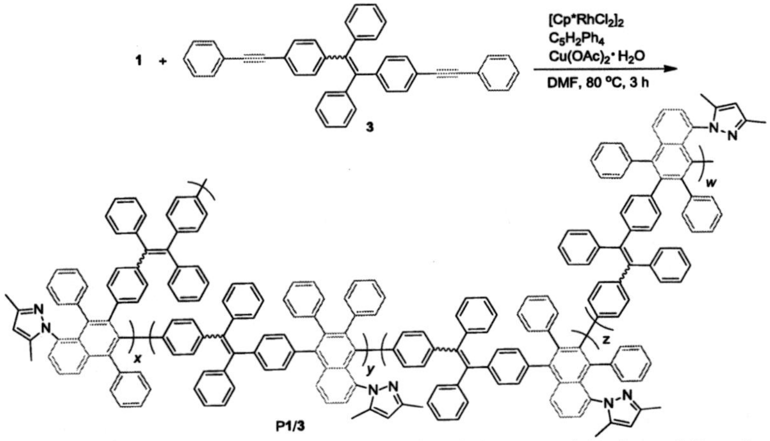

Scheme 3 Polymerization of phenylpyrazole 1 and tetraphenylethene-containing internal diyne 3.

EXPERIMENT

General information

Tetrahydrofuran (THF) was distilled under nitrogen from sodium benzophenone ketyl immediately prior to use. Dimethylformamide (DMF) was distilled over calcium hydride and stored over molecular sieves. The rhodium complex [RhCp*Cl$_2$]$_2$, named 1,2,3,4-tetramethylcyclopentadienylrhodium(III) chloride dimer was prepared according to the literature method.[18] Chemicals such as copper(II) acetate monohydrate Cu(OAc)$_2$·H$_2$O, silver 4-toluenesulfonate (AgOTs), silver trifluoromethanesulfonate (AgCF$_3$SO$_3$) and silver trifluoromethanecarboxylate (AgCF$_3$CO$_2$), and other reagents were all purchased from Aldrich and used as received without further purification.

Weight-(M_w) and number-average (M_n) molecular weights and polydispersities (M_w/M_n) of the polymers were estimated by a Waters Gel Permeation Chromatography (GPC) system equipped with a Waters 515 HPLC pump, a set of Styragel columns (HT3, HT4 and HT6; molecular weight range 10^2–10^7), a column temperature controller, a Waters 486 wavelength-tunable UV-vis detector, a Waters 2414 differential refractometer and a Waters 2475 fluorescence detector. The polymers were dissolved in THF (~1 mg/mL) and filtered through 0.45 µm PTFE syringe-type filters before being injected into the GPC system. THF was used as eluent at a flow rate of 1.0 mL/min. The column temperature was maintained at 40 °C and the working wavelength of the UV-vis detector was set at 254 nm. A set of monodispersed polystyrene standards (Waters) covering the molecular weight range of 10^3–10^7 were used for the molecular weight calibration. IR spectra were recorded on a Perkin-Elmer 16 PC FTIR spectrophotometer. ^{1}H and ^{13}C NMR spectra were measured on Bruker ARX 400 NMR spectrometers using chloroform-*d* as solvent. High resolution mass spectra (HRMS) were

5

recorded on a GCT Premier CAB 048 mass spectrometer operated in MALDI-TOF model. Thermogravimetric analyses (TGA) were conducted under nitrogen on a Perkin-Elmer TGA 7 analyzer at a heating rate of 10 °C/min. Particle sizes of the polymer aggregates in THF/water mixture were measured on a BeCoulter Delsa 440SX Zeta potential analyzer. Refractive indices were determined on a J A Woollam Variable Angle Ellipsometry System with a wavelength tunability from 300 to 1000 nm.

Monomer Synthesis

Monomer **3** was prepared according to a modified literature method[19] as stated below. To a suspended solution of 1,2-bis(4-bromophenyl)-1,2-diphenylethene (0.98 g, 2.0 mmol) in triethylamine (30 mL) and toluene (15 mL), triphenylphosphine (105 mg, 0.40 mmol), copper(I) iodide (76 mg, 0.40 mmol) and Pd(PPh$_3$)$_2$Cl$_2$ (140 mg, 0.20 mmol) were added under nitrogen. Phenylacetylene (0.66 mL, 6.0 mmol) was then injected through a septum under stirring and the reaction mixture was heated to 80 °C for 12 h. After being cooled to room temperature, the reaction mixture was dried under vacuum and extracted with CH$_2$Cl$_2$ (60 mL × 3). The organic layers were combined and dried over Na$_2$SO$_4$. After filtration and solvent evaporation, the crude product was purified by silica gel column chromatography using petroleum ether/ethyl acetate as eluent. Yellow solid; yield 80% (0.85 g). IR (KBr), v (cm^{-1}): 3027, 1594, 1495, 1440, 753, 694. ^{1}H NMR (CDCl$_3$, 400 MHz), δ (ppm): 7.50–7.47 (m, 4H), 7.32–7.24 (m, 10H), 7.14–6.99 (m, 14H). ^{13}C NMR (CDCl$_3$, 100 MHz), δ (ppm): 143.1, 143.0, 142.51, 142.46, 140.3, 131.0, 130.9, 130.8, 130.7, 130.5, 130.4, 127.7, 127.6, 127.3, 127.2, 126.3, 126.2, 122.7, 120.8, 120.6, 89.2, 89.1, 88.90, 88.86. HRMS (MALDI-TOF): m/z 532.2186 (M$^+$, calcd 532.2191).

Model reaction

To a 15 mL Schlenk tube with a three-way stopcock on the sidearm was placed [RhCp*Cl$_2$]$_2$ (2.47 mg, 0.004 mmol), C$_5$H$_2$Ph$_4$ (5.95 mg, 0.016 mmol), Cu(OAc)$_2$·H$_2$O (40 mg, 0.20 mmol), **1** (34.4 mg, 0.20 mmol) and **4** (41.6 mg, 0.20 mmol) under nitrogen. Freshly dried DMF (0.5 mL) was then injected into the tube using a hypodermic syringe. The resulting mixture was stirred at 80 °C under nitrogen for 3 h. The reaction mixture was then cooled to room temperature and extracted with CH$_2$Cl$_2$ (30 mL × 3). The organic layer was washed with water (50 mL × 3) and dried over Na$_2$SO$_4$. After purification by silica gel column chromatography using hexane/ethyl acetate mixture as eluent, a yellow solid of a mixture of **5–8** (53.3 mg, 91%) was obtained. IR (KBr), v (cm^{-1}): 1281, 1236, 1171 (C–O stretching). ^{1}H NMR (CDCl$_3$, 400 MHz), δ (ppm): 7.79–5.20 (aromatic protons), 3.75, 3.63, 3.56, 3.54, 3.52 (OCH$_3$ protons), 2.04, 2.02, 2.00 (CH$_3$ protons). ^{13}C NMR (CDCl$_3$, 100 MHz), δ (ppm): 158.0–110.7, 55.1, 55.0, 54.7, 13.2, 12.1. HRMS (MALDI-TOF): m/z 586.2661 (M$^+$, calcd 586.2620).

Polymerization

All the polymerization reactions were carried out under nitrogen atmosphere using a standard Schlenk technique. A typical procedure for the polymerization of **1** and **2(4)** is given below as an example.

To a 15 mL Schlenk tube with a three-way stopcock on the sidearm were placed [RhCp*Cl$_2$]$_2$ (2.47 mg, 0.004 mmol), C$_5$H$_2$Ph$_4$ (5.95 mg, 0.016 mmol), Cu(OAc)$_2$·H$_2$O (40 mg, 0.20 mmol), **1** (34.4 mg, 0.20 mmol) and **2(4)** (44.2 mg, 0.10 mmol) under nitrogen. Freshly distilled DMF (0.5 mL) was then injected into the tube using a hypodermic syringe. The resulting mixture was stirred at 80 °C under nitrogen for 3 h. The solution was added dropwise into 200 mL of methanol via a cotton filter under stirring. The precipitate was allowed to stand overnight and then collected by filtration. The polymer was washed with methanol and dried under vacuum at room temperature to a constant weight. Brown powder of polymer P1/2(4) was obtained in 82% yield. M_w 26300; M_w/M_n 1.50 (Table 3, no.1). IR (KBr), v (cm^{-1}): 1280, 1236, 1170, 1105, 1019 (C–O stretching). ^{1}H NMR (CDCl$_3$, 400 MHz), δ (ppm): 7.83–5.34 (aromatic protons), 3.97–3.67 (OCH$_3$ protons), 2.01–1.66 (CH$_3$ protons). ^{13}C NMR (CDCl$_3$, 100 MHz), δ (ppm): 157.5, 156.5, 156.3, 142.9–111.8, 67.4, 67.3, 67.0, 25.8, 13.3, 12.4.

   P1/2(6): Brown powder; yield 74%. M_w 27300; M_w/M_n 1.53 (Table 3, no. 2). IR (KBr), v (cm^{-1}): 1281, 1236, 1171, 1106, 1019 (C–O stretching). ^{1}H NMR (CDCl$_3$, 400 MHz), δ (ppm): 7.40–5.43 (aromatic protons), 3.92–3.70 (OCH$_3$ protons), 2.01–1.33 (CH$_3$ protons). ^{13}C NMR (CDCl$_3$, 100 MHz), δ (ppm): 157.6, 156.8, 140.4–111.9, 67.7, 67.6, 67.4, 29.1, 25.8, 13.5, 12.5.

   P1/2(8): Brown powder; yield 79%. M_w 26000; M_w/M_n 1.58 (Table 3, no. 3). IR (KBr), v (cm^{-1}): 1280, 1237, 1171, 1106, 1023 (C–O stretching). ^{1}H NMR (CDCl$_3$, 400 MHz), δ (ppm): 7.85–5.44 (aromatic protons), 3.92–3.68 (OCH$_3$ protons), 2.08–1.21 (CH$_3$ protons). ^{13}C NMR (CDCl$_3$, 100 MHz), δ (ppm): 157.6, 156.7, 156.5, 140.7–111.7, 67.8, 67.7, 67.5, 29.2, 29.1, 25.8, 13.5, 12.5, 10.3.

   P1/2(10): Brown powder; yield 77%. M_w 35700; M_w/M_n 1.66 (Table 3, no. 4). IR (KBr), v (cm^{-1}): 1282, 1241, 1174, 1107, 1028 (C–O stretching). ^{1}H NMR (CDCl$_3$, 400 MHz), δ (ppm): 7.84–5.29 (aromatic protons), 3.91–3.68 (OCH$_3$ protons), 2.07–1.21 (CH$_3$ protons). ^{13}C NMR (CDCl$_3$, 100 MHz), δ (ppm): 157.7, 156.7, 149.1–105.6, 67.9, 67.8, 67.6, 29.4, 29.1, 26.0, 25.9, 13.5, 12.3,

   P1/3: Brown powder; yield 66%. M_w 17900; M_w/M_n 1.71 (Table 3, no. 5). IR (KBr), v (cm^{-1}): 3023, 1599, 1553, 1492, 1439, 1382, 751, 696 (C–O stretching). ^{1}H NMR (CDCl$_3$, 400 MHz), δ (ppm): 7.82–6.42 (aromatic protons), 2.05–1.99 (CH$_3$ protons). ^{13}C NMR (CDCl$_3$, 100 MHz), δ (ppm): 149.3–105.8, 13.5, 12.3.

DISCUSSION
Monomer synthesis
   To develop the oxidative coupling of phenylpyrazole and internal alkyne into a versatile methodology for the construction of functional polymers, we synthesized 3,5-dimethyl-1-phenylpyrazole (1)[11] and 4,4'-(α,ω-alkylenedioxy) bis(diphenylacetylene) 2(*m*)[9] (Scheme 2) according to the literature methods. Considering that tetraphenylethene (TPE)[12] shows a phenomenon of aggregation induced emission (AIE), we also prepared 1,2-bis[4-(2-phenylethynyl)phenyl]-1,2-diphenyl-ethene 3 (Scheme 3) by palladium-catalyzed cross-coupling of 1,2-bis(4-bromophenyl)-1,2-diphenylethene[13] and phenylacetylene, whose polymerization with 1 was expected to give a polymer with a fully conjugated structure. All the monomers were characterized by standard spectroscopic methods and gave satisfactory analysis data corresponding to their structures (see data given in the Experimental Section).

Model reaction
   Before studying the polymerization of 1 with 2(*m*) or 3, we first examined whether 1 could couple with asymmetric internal alkyne 4 (Scheme 4).[9] The oxidative coupling reaction was catalyzed by [Cp*RhCl$_2$]$_2$ in the presence of Cu(OAc)$_2$·H$_2$O as oxidant and 1,2,3,4-tetraphenylcyclopenta-1,3-diene (C$_5$H$_2$Ph$_4$) as replacement ligand in dimethylformamide (DMF) at 80 °C under nitrogen, which furnished a mixture of pyrazolylnaphthalenes 5–8 in a total yield of 91% after purification. Since 5–8 possess similar physical properties, it is hard to separate them by column chromatography. They also resonate at very similar chemical shifts, thus

preventing us to determine their molar ratios by ^{1}H NMR spectroscopy. However, their high resolution mass spectrum exhibited M$^+$ peak at 586.2661, which was in well agreement with their calculated molar mass (586.2620) and hence confirmed their structures.

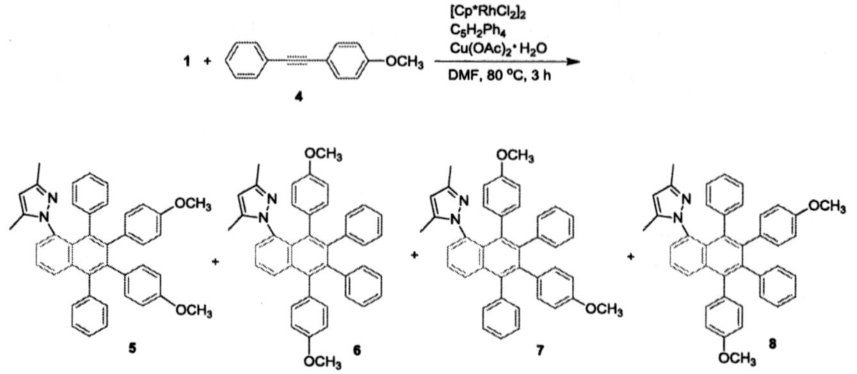

Scheme 4 Synthesis of pyrazolylnaphthalenes **5–8** by rhodium-catalyzed oxidative coupling of **1** and **4**.

Polymerization

Although **1** can be coupled with **4** efficiently under the conditions stated in Scheme 4, their suitability for the polymerization of **1** and **2**(m) or **3** remain uncertain. To search for optimum parameters, we first studied the effects of different oxidants and temperature on the oxidative polycoupling reaction using **1** and **2**(4) as monomers. In the presence of Cu(OAc)$_2$·H$_2$O, a polymer was obtained in a good yield of 82% with an M_w value of 26300 at 80 °C for 3 h (Table 1, no. 1). Temperature exerted a strong influence on the polymerization: a polymer was isolated in a much lower yield of 26% at 60 °C (Table 1, no. 2). While silver salts such as AgOTs, AgCF$_3$SO$_3$ and AgCF$_3$CO$_2$ are good oxidants to oxidize Rh^{I}X to Rh^{III}X$_3$[10] and for the polycoupling of phenylboronic acid and internal diyne,[9] we found that none of them was capable of initiating the polymerization (Table 1, nos. 3–5).

Table 1 Effects of oxidant and temperature on the polymerization of **1** and **2**(4)^a

No.	Oxidant	Temp (°C)	Yield (%)	$M_w{}^b$	$M_w/M_n{}^b$
1	Cu(OAc)$_2$·H$_2$O	80	82	26300	1.50
2	Cu(OAc)$_2$·H$_2$O	60	26	24600	1.89
3	AgOTs	80	0		
4	AgCF$_3$SO$_3$	80	0		
5	AgCF$_3$CO$_2$	80	0		

^{a} Carried out in DMF under nitrogen for 3 h in the presence of [Cp*RhCl$_2$]$_2$, C$_5$H$_2$Ph$_4$ and different oxidants. [**1**] = 0.40 M, [**2**(4)] = 0.20 M, [Rh] = 8 mM. [C$_5$H$_2$Ph$_4$] = 0.032 M and [oxidant] = 0.40 M. ^{b} Estimated by GPC in THF on the basis of a polystyrene calibration.

The influence of monomer and catalyst concentrations on the polymerization was then investigated. When the concentrations of **1** and **2(4)** were increased progressively from 0.10 M and 0.50 M to 0.40 M and 0.20 M, respectively but keeping the monomer and catalyst feeding ratios unchanged, both the isolated yield and molecular weight of the polymer were enhanced accordingly (Table 2, nos. 1–4). The polymerization of **1** and **2(4)** carried out at a molar ratio of 1:1 gave a poorer result (Table 2, no. 5). Thus, it seems that it is beneficial to perform the polymerization in an excess amount of **1** or under stoichiometric imbalance conditions.

Table 2 Concentration effect on the polymerization of **1** and **2(4)**[a]

No.	[1] (M)	[2(4)] (M)	Yield (%)	M_w^b	M_w/M_n^b
1	0.10	0.05	61	19400	2.19
2	0.15	0.075	68	22600	1.85
3	0.20	0.10	76	25100	1.70
4[c]	0.40	0.20	82	26300	1.50
5	0.20	0.20	71	18200	1.61

[a] Carried out in DMF at 80 °C under nitrogen for 3 h in the presence of [Cp*RhCl$_2$]$_2$, C$_5$H$_2$Ph$_4$ and Cu(OAc)$_2$•H$_2$O. Molar ratio of [**1**]:[**2(4)**]:[Rh]:[C$_5$H$_2$Ph$_4$]:[Cu] = 2.0:1.0:0.02:0.08:2.0. [b] Estimated by GPC in THF on the basis of a polystyrene calibration. [c] Data taken from Table 1, no. 1.

Table 3 summarizes the polymerization of other monomer pairs under optimum conditions. All the polymerization reactions proceeded smoothly, giving their corresponding polymers **P1/2(m)** and **P1/3** in good yields with high molecular weights up to 35700 and narrow distributions down to 1.5 (Table 3, nos. 1–5). Even they are constructed from aromatic rings, all the polymers show good solubility in common organic solvents, such as dichloromethane, chloroform, tetrahydrofuran (THF) and toluene, and can fabricate into tough, thin solid films by spin-coating or static-casting processes, presumably due to the irregular distribution of the repeating units of **5–8** along the polymer chains.

Table 3 Polymerization of **1** and **2(m)** or **3** under optimized conditions[a]

No.	Polymer	Yield (%)	M_w^b	M_w/M_n^b
1[c]	P1/2(4)	82	26300	1.50
2	P1/2(6)	74	27300	1.53
3	P1/2(8)	79	26000	1.58
4	P1/2(10)	77	35700	1.66
5	P1/3	66	17900	1.71

[a] Conducted in DMF at 80 °C under nitrogen for 3 h in the presence of [Cp*RhCl$_2$]$_2$, C$_5$H$_2$Ph$_4$ and Cu(OAc)$_2$•H$_2$O. [**1**] = 0.40 M, [**2(m)**] = 0.20 M, [Rh] = 8 mM, [C$_5$H$_2$Ph$_4$] = 32 mM and [Cu] = 0.40 M. [b] Estimated by GPC in THF on the basis of a polystyrene calibration. [c] Data taken from Table 1, no. 1.

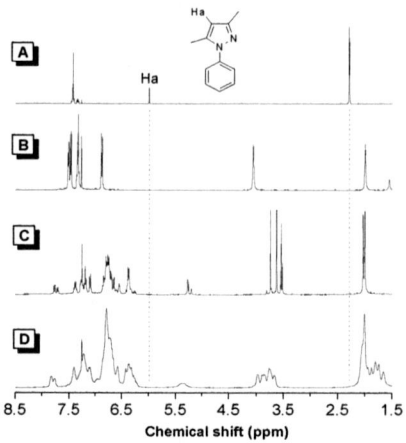

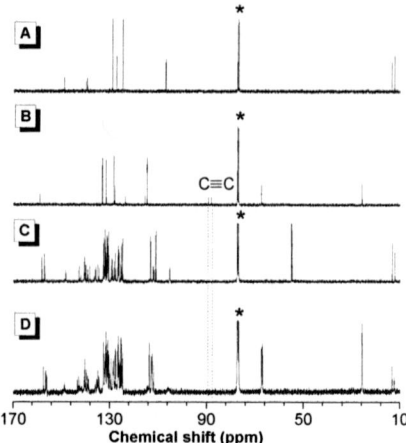

Fig. 1 ^{1}H NMR spectra of CDCl$_3$ solutions of (A) **1**, (B) **2(4)**, (C) a mixture of **5–8** and (D) P**1/2(4)** (sample taken from Table 3, no. 1).

Fig. 2 ^{13}C NMR spectra of CDCl$_3$ solutions of (A) **1**, (B) **2(4)**, (C) a mixture of **5–8** and (D) P**1/2(4)** (sample taken from Table 3, no. 1). The solvent peaks were marked with asterisks.

Structural characterization

All the polymers were characterized by standard spectroscopic methods and gave satisfactory analysis data corresponding to their structures. Fig. 1 shows the ^{1}H NMR spectra of P**1/2(4)**, **1**, **2(4)** and a mixture of model compounds **5–8**. The absorption peak at δ 5.99 in **1** was associated with the aromatic proton resonance of the pyrazole ring, which shifted upfield to δ 5.34 in the spectrum of P**1/2(4)**. The methyl proton resonances at δ 2.31 and 2.30 in **1** also shifted to δ 2.01 after the polymerization. Meanwhile, new peaks emerged at the aromatic absorption regions of δ 7.83–7.78 and 6.42–6.28 in P**1/2(4)** due to the formation of new naphthalene rings in the polymer. The spectrum of P**1/2(4)** resembled to that of **5–8** but depicted much broader absorption peaks, revealing its polymeric nature.

The ^{13}C NMR spectrum of P**1/2(4)** exhibited no resonance peaks of acetylene carbon atoms of **2(4)** at δ 89.2 and 88.9 (Fig. 2). Its spectral pattern largely resembled to that of **5–8**, confirming that the polymeric product was indeed P**1/2(4)** with a molecular structure as shown in Scheme 2.

Thermal stability

The thermal stability of P**1/2(m)** and P**1/3** was evaluated by thermogravimetric analysis (TGA). As shown in Fig. 3, all the polymers were thermally stable, losing 5% (T_d) of their weight at temperatures higher than 300 ^{o}C under nitrogen. Among them, polymer P**1/3** showed the highest thermal stability (T_d = 422 ^{o}C), which might be ascribed to its higher aromatic content.

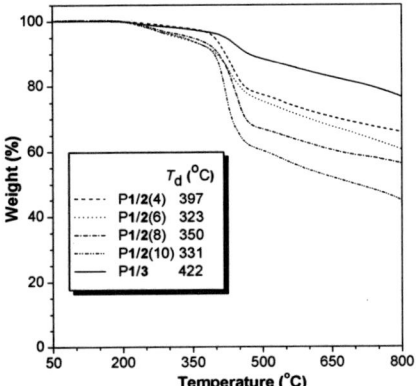

Fig. 3 TGA thermograms of P1/2(*m*) and P1/3 recorded under nitrogen at a heating rate of 10 °C/min.

Light refraction

Polymers with high refractive indices (RI or *n*) are promising candidate materials for various practical applications, including lenses, prisms, optical waveguides and holographic image recording systems.14 As P1/2(m) and P1/3 are constructed from aromatic rings, they may show high refractive indices. Indeed, light yellow transparent thin films of the polymers deposited on silica substrate showed high RI values (1.747–1.593) in a wide wavelength region of 400–1000 nm (Fig. 4).

The RI values of the polymers at 632.8 nm were all > 1.61 (Table 4), which were much higher than those of the commercially important optical plastics, such as polyacrylate (*n* = 1.492), polycarbonate (*n* = 1.581) and polystyrene (*n* = 1.587). The magnitude of the RI value at 632.8 nm was in the order of P1/3 (1.725) > P1/2(4) (1.628) > P1/2(6) (1.625) > P1/2(8) (1.617) > P1/2(10) (1.613), which was in some sense expected and was in well correlation with their aromatic content. The Abbé number (v_D) of a material is a measure of the variation or dispersion in its RI value with wavelength. The v_D values of P1/2(m) and P1/3 were in the range of 16–92, corresponding to *D* values of 0.01–0.06. Thus, the high refractivity and low optical dispersion of the polymers make them promising as optical materials.

Table 4. Refractive indices and chromatic dispersions of P1/2[a]

No.	Polymer	$n_{632.8}$	v_D	*D*
1[c]	P1/2(4)	1.6278	16.7239	0.0598
2	P1/2(6)	1.6245	18.0229	0.0555
3	P1/2(8)	1.6165	18.6389	0.0537
4	P1/2(10)	1.6128	19.6315	0.0509
5	P1/3	1.7254	91.8275	0.0109

[a] All data were taken from Fig. 4. Abbreviation: *n* = refractive index (at 632.8 nm). v_D = Abbé number = $(n_D–1)/(n_F–n_C)$, where n_D, n_F and n_C are the RI values at wavelengths of Fraunhofer D, F and C spectral lines of 589.2, 486.1 and 656.3 nm, *D* = chromatic dispersion = $1/v_D$.

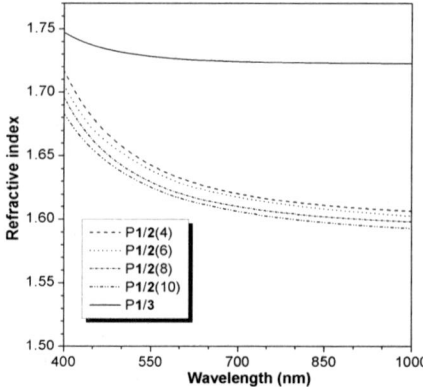

Fig. 4 Wavelength dependence of refractive index of thin films of P**1/2**(*m*) and P**1/3**.

Optical property

The UV spectrum of a diluted THF solution (10 μM) of P**1/2**(4) exhibited a broad hump centered at ~303 nm (Fig. 5). The spacer length exert little influence on the ground-state electronic transitions: the UV spectra of P**1/2**(6), P**1/2**(8) and P**1/2**(10) are practically the same to that of P**1/2**(4). On the contrary, P**1/3** absorbs at the redder region with higher intensity, thanks to the conjugated TPE unit. The absorption maximum (327 nm) occurs at much longer wavelength than that of TPE (299 nm),[12e] indicative of extensive conjugation in the polymer system owing to the electronic communication between the periphery TPE unit and the central naphthalene core.

Since TPE is an archetypical AIE luminogen, P**1/3** is thus anticipated to be AIE-active. This is indeed the case, as suggested by the photographs of its THF solution and THF/water mixtures shown in Fig. 6A. While the pure THF solution of P**1/3** emitted no light upon UV irradiation, weak green emission was observed in THF/H$_2$O mixture with 20% water content, whose intensity increases with increasing the water content.

Instead of visual observation, we also measured the photoluminescence (PL) of P**1/3** in the solution and aggregated states using a spectrofluorometer. A weak broad peak centered at ~490 nm was recorded in THF solution and THF/water mixtures with low water contents (Fig. 6B). The spectral pattern remained unchanged but the PL intensity rose gradually with an increase in the water content. From THF solution to aqueous mixture with 99% water content, the relative PL intensity increased by ~14-fold (Fig. 6C). Because P**1/3** is insoluble in water, its chains must have been aggregated in aqueous mixtures with high water contents. However, the aqueous mixture is homogeneous without precipitates even at a water content of 99%, revealing that the particles are of nano-dimension. Evidently, P**1/3** exhibits the AIE phenomenon, whose mechanism is ascribed to the restriction of intramolecular rotation of the periphery phenyl rings of the TPE unit in the aggregated state.11 Clearly, unlike conventional conjugated polymers, aggregate formation has enhanced, instead of quenched, the light emission of P**1/3**.

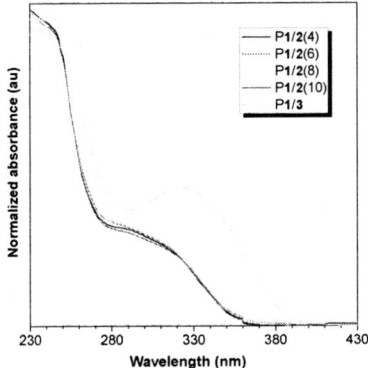

Fig. 5 Normalized UV spectra of P1/2(m) and P1/3 in THF solutions. Concentration: 10 μM.

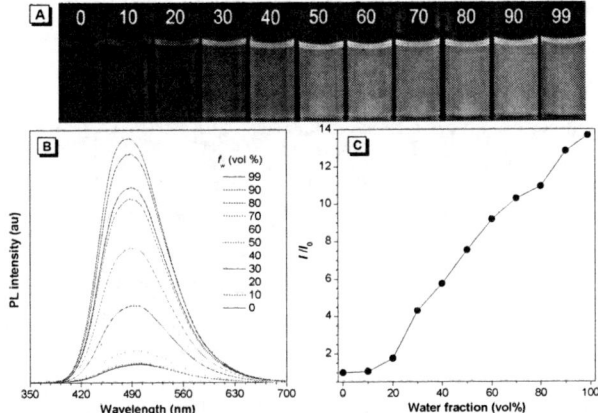

Fig. 6 (A) Photographs of P1/3 in THF/H$_2$O mixtures with different water fractions (f_w) taken under 365 nm UV irradiation from a hand-held UV lamp. (B) PL spectra of P1/3 in THF and THF/water mixtures with different f_w values. Concentration: 10 mM; excitation wavelength: 322 nm. (C) Plot of relative PL intensity (I/I_0) versus the solvent composition of THF/water mixture of P1/3.

CONCLUSIONS

In this work, we developed a new polymerization route for atom-economical synthesis of functional polymers. Oxidative polycoupling of phenylpyrazole and internal diynes was mediated by [Cp*RhCl$_2$]$_2$, C$_5$H$_2$Ph$_4$ and Cu(OAc)$_2$•H$_2$O under stoichiometric imbalance conditions in DMF for 3 h, generating poly(pyrazolylnaphthalene)s with high molecular weights in satisfactory yields. The polymers were completely soluble, film-forming and thermally stable.

Thin solid films of the polymers exhibited high refractive indices (RI = 1.747–1.593) in a wide wavelength region of 400–1000 nm. The TPE-containing poly(pyrazolylnaphthalene) was AIE-active.

ACKNOWLEDGMENTS

The work reported in this paper was partially supported by the National Basic Research Program of China (973 Program; 2013CB834701), the Research Grants Council of Hong Kong (604711, 602212, HKUST2/CRF/10 and N_HKUST620/11) and the University Grants Committee of Hong Kong (AoE/P-03/08). B. Z. Tang thanks the support of the Guangdong Innovative Research Team Program (201101C0105067115).

REFERENCES
1. (a) H. Shirakawa, *Angew. Chem. Int. Ed.*, **40**, 2575–2580(2001); (b) A. G. MacDiarmid, *Angew. Chem. Int. Ed.*, **40**, 2581–2590 (2001); (c) A. J. Heeger, *Angew. Chem. Int. Ed.*, **40**, 2591–2611 (2001).
1. (a) J. L. Liu, J. W. Y. and B. Z. Tang, *Chem. Rev.*, **109**, 5799–5867 (2009); (b) A. Qin, J. W. Y. Lam and B. Z. Tang, *Chem. Soc. Rev.*, **39**, 2522–2544 (2010); (c) J. W. Y. Lam and B. Z. Tang, *Acc. Chem. Res.*, **38**, 745–754 (2005); (d) A. Qin, J. W. Y. Lam and B. Z. Tang, *Prog. Polym. Sci.*, **37**, 182–209 (2012); (e) S. K. Choi, Y. S. Gal, S. H. Jin and H. K. Kim, *Chem. Rev.*, **100**, 1645–1681 (2000); (f) H. F. Bunz, *Acc. Chem. Res.*, **34**, 998–1010 (2001); (g) M. B. Nielsen and F. Diederich, *Chem. Rev.*, **105**, 1837–1867 (2005); (h) W. Zhang and J. S. Moore, *Angew. Chem. Int.*, **45**, 4416–4439 (2006); (i) T. J. Masuda, *Polym. Sci. Part A: Polym. Chem.*, **45**, 165–180 (2007); (j) J. Wu, W. Pisula and K. Mullen, *Chem. Rev.*, **107**, 718–747 (2007); (k) S. W. Thomas, G. D. Joly and T. Swager, *Chem. Rev.*, **107**, 1339–1386 (2007); (l) Y. Morisakia and Y. Chujo, *Prog. Polym. Sci.*, **33**, 346–364 (2008).
2. For selected examples of polycyclotrimerization, see: (a) J. B. Shi, C. J. W. Jim, F. Mahtab, J. Z. Liu, J. W. Y. Lam, H. H. Y. Sung, I. D. Williams, Y. P. Dong and B. Z. Tang, *Macromolecules*, **43**, 680–690 (2010); (b) J. Z. Liu, L. Zhang, J. W. Y. Lam, C. K. W. Jim, Y. A. Yue, R. Deng, Y. N. Hong, A. J. Qin, H. H. Y. Sung, I. D. Williams, G. C. Jia and B. Z. Tang, *Macromolecules*, **42**, 7367–7378 (2009); (c) H. Li, J. Wang, J. Z. Sun, R. Hu, A. Qin and B. Z. Tang, *Polymer Chemistry*, **3**, 1075–1083. For selected examples of Diels–Alder reactions, see: (a) H. Mukamal, F. W. Harris and J. K. Stille, *J. Polym. Sci. Part A: Polym. Chem.*, **5**, 2721–2729 (1967); (b) A. L. Rusanov, D. Y. Likhachev, P. V. Kostoglodov and N. M. Belomoina, *Polym. Sci. Ser. C*, **50**, 39–62 (2008).
3. M. Saleh, M. Baumgarten, A. Mavrinskiy, T. Schäfer and K. Müllen, *Macromolecules*, **43**, 137–143 (2010).
4. C. Li, M. Liu, N. G. Pschirer, M. Baumgarten and K. Mullen, *Chem. Rev.*, **110**, 6817–6855 (2010).
5. S. J. Toal and W. C. Trogler, *J. Mater. Chem.*, **16**, 2871–2883 (2006).
6. (a) A. M. Fraind and J. D. Tovar, *J. Phys. Chem. B*, **114**, 3104–3116 (2010); (b) J. Feng, X. Chen, Q. Han, H. Wang, P. Lu and Y. Wang, *J. Lumin.*, **131**, 2775–2783 (2011); (c) V. C. Sundar, J. Zaumseil, V. Podzoror, E. Menard, R. L. Willett, T. Someya, M. E. Gershenson and J. A. Rogers, *Science*, **303**, 1644–1646 (2004); (d) M. S. Goncalves, *Chem. Rev.*, **109**, 190–212 (2009); (e) S. Li, J. Xiang, X. Mei and C. Xu, *Tetrahedron Lett.*, **49**, 1690–1693 (2008).
7. For selected examples of polymerizations under stoichiometric imbalance conditions: (a) K. Wakabayashi, S. I. Kohama, S. Yamazaki and K. Kimura, *Macromolecules*, **41**, 1168–1174 (2008); (b) D. Zhao and K. Yue, *Macromolecules*, **41**, 4029–4036 (2008); (c) N. Nomura, K. Tsurugi and M. Okada, *Angew. Chem. Int. Ed.*, **40**, 1932–1935 (2001); (d) N. Nomura, K. Tsurugi, T. V. RajanBabu and T. Kondo, *J. Am. Chem. Soc.*, **126**, 5354–5355 (2004); (e) T. Takemura, K. Sugie, H. Nishino, S. Kawabata and T. Koizumi, *J. Polym. Sci. Part A: Polym. Chem.*, **46**, 2250–2261 (2008); (f) N. Kihara, S. Komatsu, T. Takata and T. Endo, *Macromolecules*, **32**, 4776–4783 (1999); (g) H. Iimori, Y. Shibasaki, S. Ando and M. Ueda, *Macromol. Symp.*, **199**, 23–35 (2003); (h) T. Dutta, K. B. Woody and M. D. Watson, *J. Am. Chem. Soc.*, **130**, 452–453 (2008); (i) A. R. Cruz, M. C. G. Hernandez, M. T. Guzmán-Gutiérrez, M. G. Zolotukhin, S. Fomine, S. L. Morales, H. Kricheldorf, E. S. Wilks, J. Cárdenas and M. Salmón, *Macromolecules*, **45**, 6774–6780 (2012); (j) H. R. Kricheldorf, M. G. Zolotukhin and J. Cardenas, *Macromol. Rapid Commun.*, **33**, 1814–1832 (2012).
8. M. Gao, J. W. Y. Lam, J. Li, C. Y. K. Chan, Y. Chen, N. Zhao, T. Han and B. Z. Tang, *Polymer Chemistry*, DOI: 10.1039/C2PY20758C (2013).

14

9. (*a*) N. Umeda, K. Hirano, T. Satoh, N. Shibata, H. Sato and M. Miura, *J. Org. Chem.*, **76**, 13–24 (2011); (*b*) N. Umeda, H. Tsurugi, T. Satoh and M. Miura, *Angew. Chem., Int. Ed.*, **47**, 4019–4022 (2008).
10. Z.–X. Wang and H.–L. Qin, *Green Chem.*, **6**, 90–92 (2004).
11. Y. Hong, J. W. Lam and B. Z. Tang, *Chem. Soc. Rev.*, **40**, 5361–5388 (2011); (*b*) Y. Hong, J. W. Y. Lam and B. Z. Tang, *Chem. Commun.*, 4332–4353 (2009); (*c*) R. Hu, J. W. Y. Lam, J. Liu, H. H. Y. Sung, I. D. Williams, Z. Yue, K. S. Wong, M. M. F. Yuen and B. Z. Tang, *Polym. Chem.*, **3**, 1481–1489 (2012); (*d*) H. Li, J. Wang, J. Z. Sun, R. Hu, A. Qin and B. Z. Tang, *Polym. Chem.*, **3**, 1075–1083 (2012); (e) Z. Zhao, S. Chen, J. W. Y. Lam, C. K. W. Jim, C. Y. K. Chan, Z. Wang, P. Li, C. Peng, H. S. Kwok, Y. Ma and B. Z. Tang, *J. Phys. Chem. C*, **114**, 7963–7972 (2010).
12. Y. Liu, C. M. Deng, L. Tang, A. J. Qin, R. R. Hu, J. Z. Sun and B. Z. Tang, *J. Am. Chem. Soc.*, **133**, 660–663 (2011).
13. J.–g. Liu and M. Ueda, *J. Mater. Chem.*, **19**, 8907–8919 (2009).
14. (*a*) S. W. Thomas, G. D. Joly and T. M. Swager, *Chem. Rev.*, **107**, 1339–1386 (2007); (*b*) D. Zhao and T. M. Swager, *Macromolecules*, **38**, 9377–9384 (2005).
15. (*a*) J. Liu, Y. Zhong, P. Lu, Y. Hong, J. W. Y. Lam, M. Faisal, Y. Yu, K. S. Wong and B. Z. Tang, *Polym. Chem.*, **1**, 426–429 (2010).
16. H. Sohn, M. J. Sailor, D. Magde and W. C. Trogler, *J. Am. Chem. Soc.*, **125**, 3821–3830 (2003).
17. J. W. Kang, K. Moseley and P. M. Maitlis, *J. Am. Chem. Soc.*, **91**, 5970–5977 (1969).
18. V. S. Vyas and R. Rathore, *Chem. Commun.*, **46**, 1065–1067 (2010).

Mater. Res. Soc. Symp. Proc. Vol. 1613 © 2014 Materials Research Society
DOI: 10.1557/opl.2014.153

Sequence-Controlled Vinyl Polymers by Transition Metal-Catalyzed Step-Growth and Living Radical Polymerizations

Kotaro Satoh and Masami Kamigaito

Department of Applied Chemistry, Graduate School of Engineering, Nagoya University, Nagoya, 464-8603, Japan

ABSTRACT

The metal-catalyzed step-growth radical polymerization was achieved to enable two systems for preparing tailored polymeric structures, i.e., sequence-regulated vinyl copolymer and periodically-functionalized polymer. The former is a novel strategy for preparing sequence-regulated vinyl copolymers by step-polymerization of sequence-regulated vinyl oligomers prepared from common vinyl monomers as building blocks. The later deals the simultaneous chain- and step-growth radical polymerization, which resulted in the polymers with periodic functional groups.

INTRODUCTION

The metal-catalyzed atom transfer radical addition (ATRA or Kharasch reaction) is a highly efficient carbon–carbon bond forming radical reaction [1]. Nowadays, this chemistry is widely applied to radical addition polymerizations of vinyl monomers that developed into the metal-catalyzed living radical polymerization or atom transfer radical polymerization (ATRP) and to open a new era of precision polymer synthesis [2-4].

Recently, we have found a new class of transition metal-catalyzed step-growth radical polymerization of the designed monomer, which has a reactive carbon–halogen bond and an unconjugated carbon–carbon double bond in a single molecule [5]. This reaction is also based on the metal-mediated radical addition similarly to the metal-catalyzed controlled/living radical polymerization, which is triggered by the activation and de-activation of the carbon-halogen bond derived from an initiator molecule. Whereas one halogen atom survives at the end terminal of each polymer chain in the living polymerization, the intermolecular step-growth polymerization occurs to form carbon-carbon backbone chain with inactive carbon-halogen pendants. This conceptually new radical polymerization will provide various types of linear polymers, including the equivalents of sequence-regulated vinyl copolymers from designed monomers.

In the present paper, the step-growth radical polymerization was evolved into the two systems for preparing tailored polymeric structures, i.e., sequence-regulated vinyl copolymer and periodically-functionalized polymer. The former is a novel strategy for preparing sequence-regulated vinyl copolymers by step-polymerization of sequence-regulated vinyl oligomers prepared from common vinyl monomers as building blocks, such as styrene, acrylate, acrylamide, acrylonitrile, and vinyl chloride [6,7]. In the later part, the simultaneous chain- and step-growth radical polymerization was investigated by copolymerizing step-growth monomers with common vinyl monomers, i.e., the step-polymerization was simultaneously compatibilized with the metal-catalyzed atom transfer radical polymerization, which resulted in the polymers with periodic functional moieties [8,9].

EXPERIMENTAL DETAILS

Materials
RuCl$_2$(PPh$_3$)$_3$ (Wako), FeCl$_2$ (Aldrich; 99.99%), CuCl (Aldrich; 99.99%), and n-Bu$_3$P (KANTO ; > 98%) were used as received and handled in a glove-box (VAC Nexus) under a moisture- and oxygen-free argon atmosphere (O$_2$< 1 ppm). PMDETA (Tokyo Kasei; > 98%) was distilled over calcium hydride before use. All other reagents were purified by usual methods.

Step-Growth Radical Polymerization
Polymerization was carried out under dry nitrogen in baked glass tubes equipped with a three-way stopcock. A typical example for the polymerization procedure with the iron catalyst is given below. To a suspension of FeCl$_2$ (50.7 mg, 0.40 mmol) in toluene (1.27 mL) was added n-Bu$_3$P (0.20 mL, 0.80 mmol), and the mixture kept stirred for 24 h at 80 °C to give a homogeneous solution of the FeCl$_2$(n-Bu$_3$P)$_2$ complex. After the solution was cooled to the room temperature, the monomer (16.0 mmol) was added. The solution was evenly charged in 8 glass tubes and the tubes were sealed by flame under nitrogen atmosphere. The tubes were immersed in thermostatic oil bath at 100 °C. In predetermined intervals, the polymerization was terminated by cooling the reaction mixtures to –78 °C. Monomer conversion was determined from the concentration of residual monomer measured by gas chromatography with toluene as an internal standard.

Measurements
Monomer conversion was determined from the concentration of residual monomer measured by gas chromatography with toluene as an internal standard under He gas flow or ^{1}H NMR spectroscopy with the integ.^{1}H NMR spectra were recorded in CDCl$_3$ at 25 °C, operating at 400 MHz. The number-average molecular weight (M_n) and weight-average molecular weight (M_w) of the product polymers were determined by size-exclusion chromatography (SEC) in THF at 40 °C. The columns were calibrated against 7 standard poly(MMA) or 11 standard polystyrene samples. MALDI-TOF-MS spectra were measured with dithranol (1,8,9-anthracenetriol) as the ionizing matrix and sodium trifluoroacetate as the ion source.

DISCUSSION

Sequence-Regulated Vinyl Copolymers

We first attempted to investigate the reaction of the monomers (1-8) prepared from common vinyl monomers for the ABC- or ABCC-ordered copolymers. The monomers (1-7) have the ABC-ordered sequence, where A, B, and C represent the equivalent for vinyl chloride, styrene derivatives, and acrylates, respectively (Scheme 1) [6].

Prior to polymerizations, we examined the synthesis of the compounds with a carbon–halogen and a carbon–carbon double in one molecule (1–8), which would be the monomers for the polyaddition as well as model compounds for the sequence-united polymer chains. All the monomers were prepared by the Kharasch addition reactions between common olefins and various dichloroacetates by a ruthenium complex such as RuCl$_2$(PPh$_3$)$_3$ and RuCp*Cl(PPh$_3$)$_2$, and the subsequent allylation with allyltrimethylsilane by TiCl$_4$ led to form the monomers in good yields (>90%). Especially for the synthesis of **8**, the Kharasch 1:1 addition reaction was conducted twice for methyl acrylate and styrene, in this order.

The step-growth radical polymerizations of **1–8** were then examined with a series of transition metal complexes in toluene at 100 °C. The monomers were smoothly consumed and the conversion reached over 90% with the RuCl$_2$(PPh$_3$)$_3$, FeCl$_2$/Pn-Bu$_3$, and CuCl/PMDETA systems. Although the isolated yields of the ABC-ordered polymers (**1–7**) were relatively low even for the indene-based monomers (**6** and **7**), the products were actually polymers obtained from the sequence-ordered monomers. The low yields of the polymers were probably due to the unwanted cyclization of the monomer to form a five-membered ring at the first stage of the reaction. On the other hand, the step-growth radical polymerization of **8** with the repeating four vinyl monomer units proceeded smoothly without significant amount of the unwanted cyclization, which would give an ABCC-sequenced polymer.

The structures of the polymers thus obtained were then analyzed by ^{1}H NMR and matrix-assisted laser-desorption-ionization time-of-flight mass (MALDI-TOF-MS) spectroscopy. In the ^{1}H NMR spectra, a series of sharp peaks assigned to the double bond and the active C–Cl bonds at the chain ends were observed in addition to the broad and relatively large peaks of the sequence-regulated main chain of the polymers. The MALDI-TOF-MS spectra consisted of a series of peaks each separated by the formula weight of polymerizations proceed via an intermolecular step-polymerization to afford the equivalents of well-sequenced copolymers.

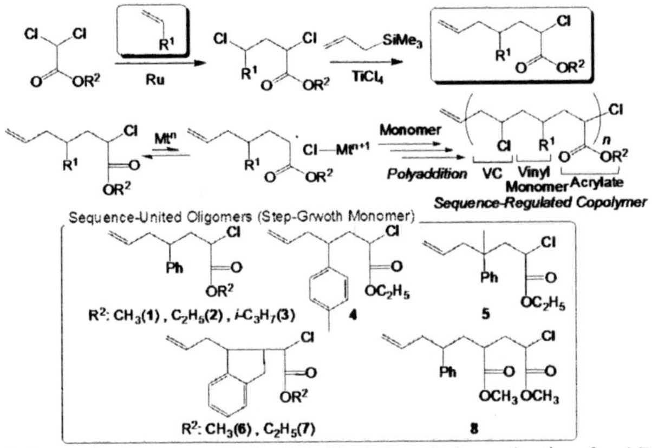

Scheme 1. Sequence-United Radical Step-Polymerization for ABC- or ABCC-Sequence-Regulated Copolymers.

Simultaneous Chain- and Step-Growth Radical Polymerization

The step-growth radical polymerization is based on the metal-mediated radical addition as well as the metal-catalyzed controlled/living radical polymerization, which is triggered by the activation and de-activation of the carbon-halogen bond derived from an initiator molecule.

Whereas one halogen atom survives at the end terminal of each polymer chain in the living polymerization, the step-growth polymerization occurs to form carbon-carbon backbone chain with inactive carbon-halogen pendants but another halogen atom still survives at the end terminal similarly to the living polymerization. We thus investigated unprecedented simultaneous chain- and step-growth polymerization by combining these two polymerizations for various common conjugated vinyl monomers and unconjugated olefins with ester- or amide-linkages (Scheme 2) [8].

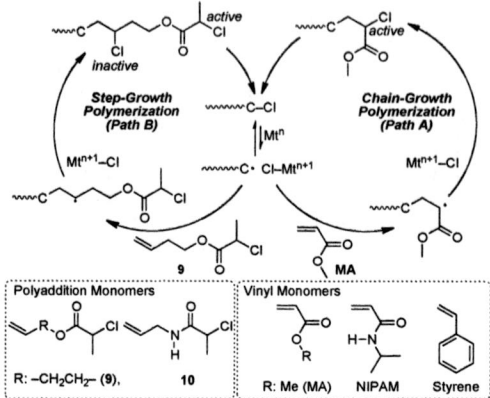

Scheme 2. Transition-Metal Catalyzed Simultaneous Chain- and Step-Growth Radical Polymerization.

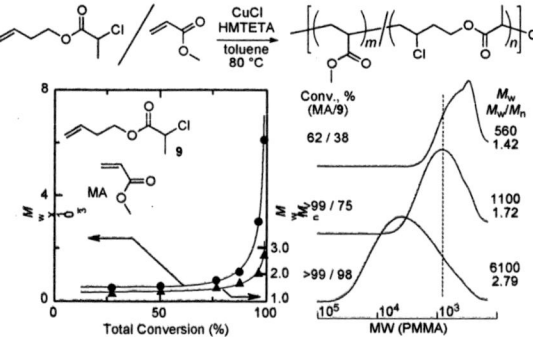

Figure 1. Simultaneous radical chain- and step-growth polymerization of MA and **9** with CuCl/HMTETA in toluene at 80 °C: $[MA]_0 = 2.0$ M; $[9]_0 = 2.0$ M; $[CuCl]_0 = 100$ mM; $[HMTETA]_0 = 100$ mM.

The simultaneous polymerizations of a common vinyl monomer, such as methyl acrylate (MA: for chain-growth polymerization), and a designed ester-linked monomer (**9**: for step-growth polymerization) were investigated with various transition-metal complexes. Especially, with the CuCl/1,1,4,7,10,10- hexamethyltriethylenetetramine (HMTETA) system, the reaction ($[MA]_0/[9]_0 = 1/1$) reached high monomer conversions (>99% for MA and 98% for 9) to give relatively high molecular weight polymers ($M_w = 6000$) (Figure 1). The ^{1}H NMR and MALDI-TOF-MS analyses revealed that the CuCl/HMTETA and RuCp*Cl(PPh$_3$)$_2$ systems induced a cross-propagation of MA and 9 with random sequences via simultaneous chain- and step-growth copolymerizations and resulted in the linear copolymers without undesirable chain-vinyl copolymerization of 9.

Further investigation was directed to the intermolecular radical step-polymerization between α,ω-difunctional compounds (AA- and BB-monomers), one of which has two carbon–carbon double bonds and the other has two carbon–halogen bonds in one molecule. The step-polymerization between two monomers also generated the various equivalents of sequence-regulated vinyl copolymers not only by monomer designing but also by the combinations of the monomers [7]. The AA- and BB-type step growth polymerization could also be combined with the living chain-polymerization to achieve the simultaneous polymerization, which resulted in the polymers with various periodic functional moeities originated from AA- and BB-monomers.

CONCLUSIONS

The metal-catalyzed intermolecular radical step-polymerization successfully proceeded for the designed monomers to provide novel polymer structures including the equivalents of sequence-regulated vinyl copolymers. This method will provide further function of the synthetic vinyl polymers by conducting various functional groups periodically and building the higher-ordered structure in a certain polymer chain.

ACKNOWLEDGMENTS

This work was supported in part by a Grant-in-Aid for Scientific Research on Innovative Areas "Fusion Materials (Creative Development of Materials and Exploration of their Function through Molecular Control; Area No. 2203)" (No. 23107515) for K.S. from the Ministry of Education, Culture, Sports, Science and Technology, Japan.

REFERENCES

1. M. S.Kharasch, E. V.Jensen,W. H. Urry, *Science* **102**, 128 (1945).
2. M. Kato, M. Kamigaito, M. Sawamoto, T. Higashimura, *Macromolecules* **28**, 1721 (1995).
3. J.-S. Wang, K. Matyjaszewski, *J. Am. Chem. Soc.* **117**, 5614 (1995).
4. K.Satoh, M.Kamigaito, M. Sawamoto, "Transition Metal Complexes for Metal-Catalyzed Atom Transfer Controlled/Living Radical Polymerization," *Polymer Science: A Comprehensive Reference*, ed. Matyjaszewski, K. Moeller, M. (Elsevier, 2012) Amsterdam, Netherlands,vol. 3, pp. 429-461 (2012).
5. K. Satoh, M. Mizutani, M. Kamigaito, *Chem. Commun.* **12**, 1260 (2007).
6. K. Satoh, S. Ozawa, M. Mizutani, K. Nagai, M. Kamigaito, *Nat. Commun.*, 1: 6, doi: 10.1038/ncomms1004 (2010).
7. K. Satoh, T. Abe, M. Kamigaito, *ACS. Symp. Ser* .**1100**, 133 (2012).
8. M. Mizutani, K. Satoh, M. Kamigaito, *J. Am. Chem. Soc.* **132**, 7498 (2010).
9. M. Mizutani, K. Satoh, M. Kamigaito, *Macromolecules* **4**, 2382 (2011).

Mater. Res. Soc. Symp. Proc. Vol. 1613 © 2014 Materials Research Society
DOI: 10.1557/opl.2014.154

Synthesis of Arborescent Polymers by "Click" Grafting

Toufic Aridi and Mario Gauthier

Department of Chemistry, Institute for Polymer Research, University of Waterloo, 200 University Avenue West, Waterloo, ON N2L 3G1, Canada

ABSTRACT

A novel method was developed for the preparation of arborescent (dendritic graft) polymers, by successive grafting reactions of linear chain segments using alkyne-azide "click" chemistry coupling. A linear polystyrene substrate was thus randomly functionalized with acetylene functionalities, by acetylation and further reaction with propargyl bromide in the presence of potassium hydroxide and 18-crown-6 in toluene. The anionic polymerization of styrene was achieved with 6-*tert*-butyldimethylsiloxy-hexyllithium to obtain polystyrene with a protected hydroxyl chain end. Deprotection of the hydroxyl group, followed by conversion into tosyl and azide functionalities yielded the material serving as side chains in the grafting reactions. Coupling of the azide-terminated side chains with the acetylene-functionalized substrate in the presence of a Cu(I) catalyst proceeded in up to 93% yield. Additional cycles of substrate functionalization and side chain coupling led to arborescent polymers of generations G1 and G2, with low polydispersity indices ($M_w/M_n \approx 1.1$), in 60-84% yield. These polymers are characterized by a very compact structure, and molecular weights increasing geometrically over successive generations. A similar methodology was also shown to work for the synthesis of arborescent polybutadiene systems, using azide-functionalized substrates and alkyne-terminated side chains. The coupling reaction proceeded in up to 76% yield under optimized conditions for these systems.

INTRODUCTION

Dendritic polymers have attracted attention due to their compact structure and unusual properties which make them potentially useful for multiple applications. Over the past two decades significant efforts have been devoted to the synthesis of dendritic molecules, which can be subdivided into three main classes, namely dendrimers, hyperbranched polymers, and arborescent polymers [1]. All dendritic molecules are characterized by a multi-level branched architecture, but arborescent polymers are distinguished from dendrimers and hyperbranched polymers by their assembly from polymeric building blocks of uniform size rather than monomers, so that very high molecular weights are attained in a few synthetic steps.

The development of living/controlled polymerization techniques has enabled the synthesis of a wide range of controlled structure polymers. Well-known reactions in organic chemistry have also been adapted to polymer chemistry and have yielded promising results to build novel structures by coupling preformed polymer chains [2]. The combination of state-of-the-art living/controlled polymer chemistry techniques with the best coupling procedures known in organic chemistry thus allows the preparation of an even wider range of materials. In this respect the concept of click chemistry, as defined by Sharpless [3-5], appears ideally suited to coupling preformed polymer chains into more complex chain architectures. Sharpless and co-workers identified a number of reactions that meet the criteria for "click" chemistry, arguably the most powerful of which is the Cu (I)-catalyzed variant of the Huisgen 1,3-dipolar cycloaddition of azides and alkynes to afford 1,2,3-triazoles [3].

In the present investigation, grafting procedures based on the azide–alkyne 1,3-dipolar cycloaddition reaction were applied to the synthesis of well-defined polystyrene and polybutadiene arborescent architectures.

EXPERIMENTAL DETAILS

Cyclohexane and toluene (reagent grade) were purified by refluxing with oligostyryllithium under dry N_2 atmosphere. Tetrahydrofuran (THF; reagent grade) was purified by distillation from sodium-benzophenone ketyl under N_2. The solvents were introduced directly from the stills into the polymerization reactor and reaction setups through polytetrafluoroethylene (PTFE) tubing. Styrene (99%) was purified by stirring with CaH_2 and distillation at reduced pressure. It was stored under N_2 at -20 °C until a second purification step with phenylmagnesium chloride (2 M in THF; 1 mL for 10 mL of monomer) immediately before polymerization. Butadiene (99%) was purified by stirring with n-BuLi (2 M in hexane; 1 mL for 10 mL of monomer) for 30 min at -30 °C and condensation to an ampoule under vacuum. The monomer was diluted by condensing an equal volume of dry THF or cyclohexane under vacuum, and the ampoule was stored at -20 °C until it was used. All the other reagents were used as received from the suppliers.

Anionic polymerization of the monomers was carried out under N_2 atmosphere initiated either by sec-BuLi (linear substrates) or by 6-$tert$-butyldimethylsiloxylhexyllithium (TBDMSLi; materials serving as side chains).

Acetylene functionalities were randomly introduced on the linear polystyrene substrate by acetylation [6], followed by reaction with propargyl bromide in the presence of potassium hydroxide and 18-crown-6 ether [7]. The $tert$-butyldimethylsiloxyl chain ends of the polystyrene side chains initiated with TBDMSLi were deprotected by treatment with tetrabutylammonium fluoride (TBAF) [8]. The hydroxyl chain ends of the polymer were then converted to a tosylate [9-11], which was displaced by sodium azide in N,N-dimethylformamide (DMF) [10, 12] to generate the azide-terminated side chains.

Coupling of the linear alkyne-functionalized polystyrene substrate with the azide-terminated side chains, to obtain a generation zero (G0) or comb-branchedpolystyrene structure, was carried out at room temperature with a 1:1 ratio of acetylene to azide functionalities in degassed DMF, using 1 eq each ofN,N,N',N'',N''-pentamethyldiethylenetriamine (PMDETA) and Cu(I)Br to catalyze the reaction. The crude G0 product was purified by precipitation fraction from toluene/methanol mixtures to remove the unreacted linear side chain contaminant. Further cycles of substrate alkynylation and "click" grafting reactions under similar conditions yielded G1 and G2 arborescent polystyrenes.

The synthesis of the analogous arborescent polybutadiene structures was accomplished starting from the linear polybutadiene substrate, randomly functionalized with azide coupling sites, and alkyne-terminated polybutadiene side chains. The azidation procedure involved a sequence of epoxidation [13], reduction [14] and tosylation reactions, followed by displacement of the tosylate with sodium azide. Polybutadiene side chains with acetylene chains ends were obtained from the corresponding TBDMSLi-initiated side chains, by deprotection with TBAF, deprotonation with NaH, and coupling with propargyl bromide [15].

Coupling of the linear azide-functionalized polybutadiene substrate with the alkyne-terminated side chains, to obtain a G0 polybutadiene structure, was accomplished with a 1:1 ratio of acetylene to azide functionalities in degassed 80:20 toluene/DMF, using 5 eq each of N,N,N',N'',N''-pentamethyldiethylenetriamine (PMDETA) and Cu(I)Br to catalyze the coupling reaction. The crude G0 product was purified by precipitation fraction from hexane/2-propanol mixtures to remove the unreacted linear side chain contaminant. Further cycles of substrate azidation and "click" grafting reactions under the same conditions yielded G1 and G2 arborescent polybutadiene structures.

The characterization of the polymers obtained was achieved at the different steps of the reaction on a Viscotek GPC max size exclusion chromatography (SEC) instrument with a VE 2001 GPC Solvent/sample Module. The GPC max unit was equipped with a Viscotek TDA 30 detector, a Triple Detector Array, a Viscotek UV 2600 detector, and three PolyAnalytik organic mixed bed columns with overall range polystyrene molecular weight range capability from 10^3 to 10^7. The polymers were analyzed in THF at a flow rate of 1 mL/min. ^{1}H NMR analysis of all the polymer samples was carried out in CDCl$_3$ on a Bruker Avance 300 MHz NMR instrument.

DISCUSSION

Synthesis of the Polystyrene Substrates and Side Chains

The strategy used for the random functionalization of the polystyrene substrates with acetylene functionalities is described in Scheme 1, whereas the introduction of azide chain end is described in Scheme 2.

The synthesis of the linear substrate started with the polymerization of styrene initiated by sec-BuLi, to obtain a polymer with a number-average molecular weight M_n = 5300 and a narrow molecular weight distribution (M_w/M_n = 1.05). The polymer was acetylated (24 mole %), subsequently reacted with KOH in toluene in the presence of 18-crown-6 ether, and quenched with propargyl bromide. The polymer obtained had an acetylenation level of 23 mole % (^{1}H NMR resonance at 2.8 ppm) and 8 mole% of residual acetyl groups (2.55 ppm), which can be explained by double substitution of a portion of the acetyl functionalities under the reaction conditions used [7].

The synthesis and functionalization of the side chains was achieved by initiating the polymerization of styrene with TBDMSLi, to obtain polystyrene chains containing a protected hydroxyl end group (M_n = 5200, M_w/M_n = 1.09). Deprotection of the hydroxyl functionality with TBAF, followed by conversion into a tosyl group and displacement with sodium azide yielded side chains with an azide chain end. Complete deprotection of the tert-butyldimethylsilyl group was confirmed by ^{1}H NMR spectroscopy analysis, by the disappearance of the two resonances at 0 and 0.8 ppm for the dimethylsilyl and tert-butyl protons, respectively. The methylene protons adjacent to the hydroxyl group still appeared at 3.56 ppm, and the integration of the peak was consistent before and after deprotection. Further derivatization into the tosyl functionality produced a shift in the resonance from 3.56 to 4.1 ppm. The tosyl to azide conversion could also be monitored by ^{1}H NMR analysis, through the shift in the signal at 4.1 ppm to 3.1 ppm for the methylene protons adjacent to the azide functionality.

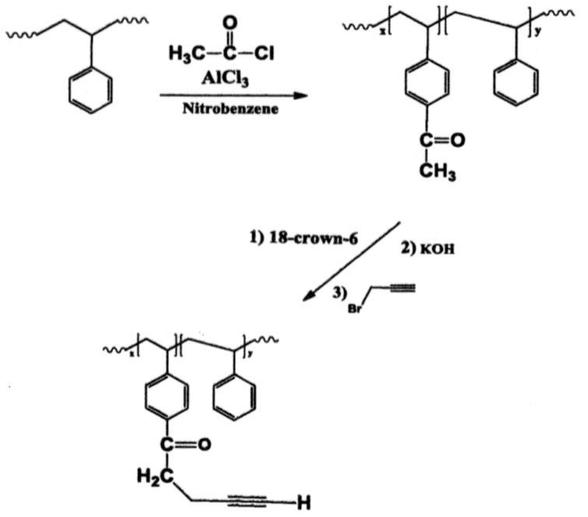

Scheme 1. Synthesis and acetylenation of linear polystyrene substrate.

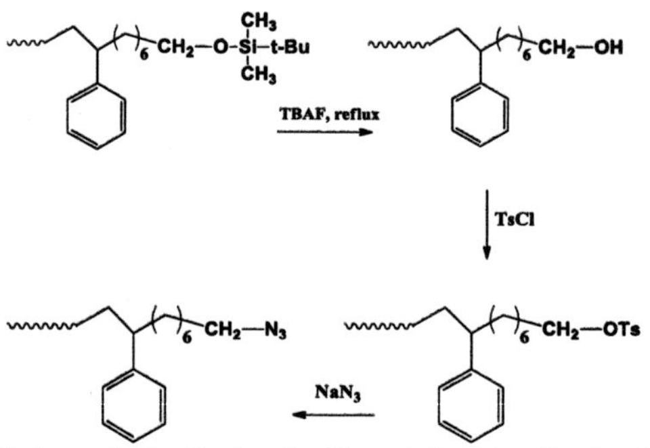

Scheme 2. Synthesis of azide end-functionalized polystyrene side chains.

Coupling of Azide-terminated Polystyrene with Acetylenated Polystyrene

The reaction conditions were optimized by exploring different solvents, amine catalysts, and reaction temperatures for the grafting reaction. Optimal results were obtained in pure DMF, using PMDETA and CuBr at room temperature, when grafting the side chains onto linear and G0 substrates. The conditions were slightly adjusted by increasing the temperature to 45°C and adding 5% of 2,6-di-*tert*-butyl-4-methylphenol (BHT, a radical scavenger) with respect to alkyne/azide groups when grafting onto G1 substrates. All the reactions were purposely performed with a 1:1 stoichiometric ratio between the azide and acetylene functionalities, to monitor the efficiency of the coupling process. Coupling of azide-terminated side chains (M_n= 5200, M_w/M_n= 1.09) with the acetylenated linear substrate (M_n= 5300, M_w/M_n= 1.08) proceeded

26

with 94% grafting yield (defined as the fraction of side chains becoming attached to the substrate). Coupling of the azide-terminated side chains with the G0 (M_n= 52000, M_w/M_n= 1.09) and G1(M_n= 4.34 ×10^5, M_w/M_n= 1.12) acetylenated substrates proceeded with 84 % and 60 % grafting yield, respectively.

It can be noticed that the grafting yield decreased as the generation number of the substrate increased, which is attributed to the fact that the G0 and G1 substrates are bulky and have an increasingly congested structure, hindering the accessibility of the alkyne groups on the substrates to the azide end groups. The SEC traces obtained for the purified graft polymers of successive generations are provided in Figure 1. It is clear that there is an increase in the size of the molecules with the generation number, as the peaks shift to the left (higher molecular weight region) of the SEC curves, while the breadth of the peaks remains relatively constant.

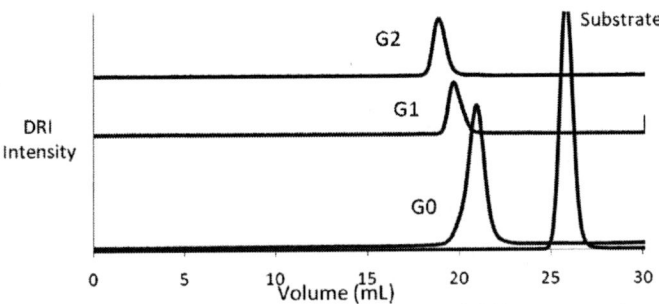

Figure 1. SEC traces for arborescent polystyrenes of different generations obtained by "click" grafting.

Characterization of the Purified Arborescent Polystyrenes

The analysis results obtained for the graft polymers are summarized in Table I. The molecular weight and the branching functionality of the graft polymers increased roughly geometrically with the generation number (Table I). The number-average branching functionality of the polymers, defined as the number of chains added in the last grafting reaction, was calculated according to the equation

$$f_n = \frac{M_n(G) - M_n(G-1)}{M_n^{br}}$$

where $M_n(G)$, $M_n(G - 1)$, and M_n^{br} are the absolute number-average molecular weight of graft polymers of generation G, of the preceding generation and of the side chains, respectively. The sample nomenclature used in Table I identifies the generation number of the substrate as well as the molecular weight of the side chains grafted in the last reaction. For example, G0PS-PS5 refers to a G0 polystyrene substrate grafted with $M_n \approx$ 5000 side chains. These results compare very well with previous reports on the synthesis of arborescent polystyrenes using anionic grafting onto chloromethylated [16] and acetylated [6] substrates. Anionic grafting of $M_n \approx$ 5000 side chains onto the linear substrates proceeded in up to 95% yield, which is essentially identical to the 94% yield obtained by "click" coupling. In terms of grafting yield, the synthesis of the higher generations (G1, G2) by "click" coupling was also comparable to anionic grafting; thus the new approach developed is an efficient method to synthesize arborescent polystyrenes. One clear advantage of this technique, however, is that anionic polymerization is only required to synthesize the building blocks (linear substrate and side chains) serving in the synthesis, and not at every grafting step serving in the synthesis of the successive generations.

Table I. Characteristics of arborescent polystyrenes of different generations.

	Side chains		Graft polymers			
	M_n^{a}	M_w/M_n^{a}	yield(%)^{b}	M_n^{a}	M_w/M_n^{a}	f_n
PS-PS5	5200	1.09	94	5.2×10^{4}	1.09	9
G0PS-PS5	5200	1.09	84	4.34×10^{5}	1.1	74
G1PS-PS5	5200	1.09	60	2.82×10^{6}	1.1	458

^{a}Absolute values determined from SEC-MALLS analysis, ^{b}fraction of side chains generated attached to the substrate.

Synthesis of the Polybutadiene Substrates and Side Chains

The strategy for the random functionalization of the polybutadiene substrates with azide functionalities is described in Scheme 3, whereas the introduction of acetylene chain end is described in Scheme 4.

The synthesis of the linear polybutadiene substrate was achieved in toluene with *sec*-BuLi, to afford a polymer with M_n= 5100 and a low polydispersity index (M_w/M_n= 1.04), having a chain microstructure with 87% of 1,4- and 13% of 1,2-butadiene units. This polymer was first subjected to epoxidation with 3-chloroperbenzoic acid, to produce a substrate with 26 mole% of epoxidized butadiene units (^{1}H NMR resonances at 2.65 and 2.95 ppm). The polymer was further reacted with LiAlH$_4$ (4 eq with respect to the epoxide groups) before workup with methanol/HCl to isolate the hydroxylated polymer (complete conversion of the epoxide groups, as determined by NMR analysis, and appearance of a resonance at 3.56 ppm). Tosylation with 10 eq of tosyl chloride, followed by displacement with 5 eq of sodium azide with respect to the tosyl groups yielded the azidated linear substrate (26 mole% substitutionby ^{1}H NMR analysis, resonance at 3.2 ppm).

The hydroxyl end-functionalized polybutadiene side chains were obtained as described above for polystyrene, by initiating the polymerization of butadiene with TBDMSLi and deprotecting the hydroxyl chain ends by treatment with TBAF. The end-functionalized polymer had M_n= 5200 and M_w/M_n= 1.09. The side chains were further treated with 5 eq of NaH in tetrahydrofuran, before adding 5 eq of propargyl bromide to achieve the quantitative conversion to acetylene chain ends. The ^{1}H NMR spectrum obtained for the acetylenated polymer still had a resonance at 3.6 ppm for the methylene protons adjacent to the ether linkage formed, but also a new resonance at 4.1 ppm for the methylene protons in the propargyl group. The integration of the peaks for the two methylene signals provided the same result, thus confirming full conversion of the hydroxyl to acetylene functionalities.

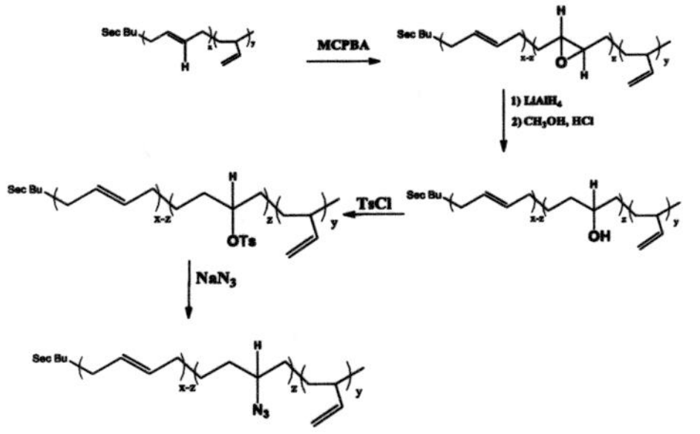

Scheme 3. Synthesis and azide functionalization of polybutadiene substrate.

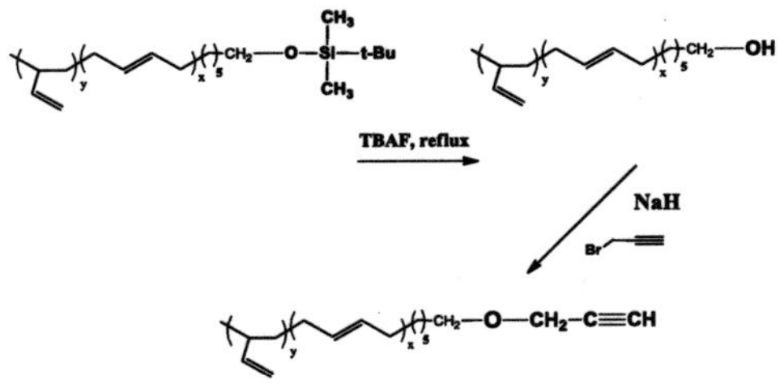

Scheme 4. Synthesis and alkyne-end functionalization of polybutadiene side chains.

Coupling of Alkyne-terminated Polybutadiene with Azidated Polybutadiene

The conditions used for the grafting reaction were optimized, as in the case of polystyrene "click" grafting, by maximizing the grafting yield while using a 1:1 stoichiometry of alkyne to azide functional groups in the reaction. Variations in the solvent, the temperature, and the amine ligand used showed that the yield was maximized (76%) in the synthesis of the G0 macromolecules in a toluene/DMF (80/20) solvent mixture, with PMDETA as ligand, and by performing the reaction at 50 °C in the presence of 5% BHT with respect to the alkyne/azide groups. Due to the high unsaturation level of the components involved in the reaction, the addition of BHT was found to be crucial to avoid significant broadening of the molecular weight distribution for reactions carried out at 50 °C, even in the synthesis of the G0 polymer. Further coupling of the acetylene-terminated side chains with the azidated G0 (M_n= 84000,M_w/M_n = 1.12) and G1 (M_n = 6.2×10^5,M_w/M_n= 1.14) substrates proceeded in 59% and 40% yield, respectively. The decreasing trend in the grafting yield is analogous to the one observed within the arborescent polystyrene series, albeit the absolute yields attained are lower in this case. This decrease is attributed to the lower polarity of the toluene/DMF mixture used in the reaction. Indeed, the low polarity of polybutadiene precluded using more than 20% of DMF by volume in the reaction mixture to avoid precipitation of the polymer.

Characterization of the Purified Arborescent Polybutadienes

The characterization data obtained by SEC analysis of the polybutadiene samples are summarized in Table II. Similarly to Table I, the molecular weight and the branching functionality of the products increased roughly geometrically over successive generations, but the absolute values of each are actually higher than for the arborescent polystyrene systems in spite of the lower grafting yields attained. While $M_n = 5000$ side chains were used as building blocks in both cases, the degree of polymerization of the polybutadiene chains ($X_n = 92$) is significantly larger than for the polystyrene chains ($X_n = 48$), which explains the higher molecular weights and branching functionalities attained in the polybutadiene systems. The synthesis of the arborescent polybutadiene structures by the "click" grafting technique developed is thus viewed as successful, in spite of the somewhat lower grafting yield attained in comparison with the arborescent polystyrene systems. It may be possible to compensate for the lower grafting yields if desired, by increasing the stoichiometric ratio of azide to alkyne groups used, i.e. by adding an excess of azidated polybutadiene substrate in the reactions.

Table II. Characteristics of arborescent polybutadienes of different generations.

	Side chains		Graft polymers			
	$M_n{}^a$	$M_w/M_n{}^a$	yield(%)^b	$M_n{}^a$	$M_w/M_n{}^a$	f_n
PBD-PBD5	5200	1.09	76	8.4×10^4	1.12	15
G0PBD-PBD5	5200	1.09	47	6.2×10^5	1.14	103
G1PBD-PBD5	5200	1.09	40	3.4×10^6	1.12	534

^{a}Absolute values determined from SEC-MALLS, ^{b}fraction of side chains generated attached to the substrate.

CONCLUSIONS

Arborescent polystyrenes and polybutadienes of generations G0-G3, with molecular weights and branching functionalities reaching over 3×10^6 and 500, respectively, were successfully synthesized for the first time by "click" chemistry coupling. The results presented show that "click" grafting can be very useful in the synthesis of arborescent polymers since, in contrast to the anionic grafting techniques previously used, it does not depend on the presence of "living" macroanions at every step of the reaction. This approach also provides a new dimension to the synthesis of these complex architectures, while avoiding the tedious anionic grafting procedures. The side chain materials and the substrates are both stable compounds, which can be stored and used at some later time without problems. This methodology should also be applicable to the synthesis of a wider range of arborescent copolymer structures, as the success of anionic grafting reactions was limited when combining lower reactivity anions and/or substrates in the coupling reaction [17].

ACKNOWLEDGMENTS

The authors thank the Natural Sciences and Engineering Research Council of Canada (NSERC) and LANXESS Inc. for their financial support of this work.

REFERENCES

1. T. Aridi and M. Gauthier in *Complex Macromolecular Architectures: Synthesis, Characterization, and Self-Assembly*, edited by N. Hadjichristidis, A. Hirao, Y. Tezuka, F. DuPrez. (Wiley, New York, 2011, Chapter 6).
2. C. J. Hawker and K. L. Wooley, *Science* **309**, 1200 (2005).
3. H. C. Kolb, M. G. Finn and K. B. Sharpless, *Angew. Chem, Int. Ed.* **40**, 2004 (2001).
4. H. C. Kolb and K. B. Sharpless, *Drug Discovery Today* **8**, 1128 (2003).
5. R. Huisgen, in *1,3-Dipolar Cycloadditional Chemistry*, edited by A. Padwa. (Wiley, New York, 1984).
6. J. Li and M. Gauthier, *Macromolecules* **34**, 8918 (2001).
7. E. Abele, R. Abele, Y. Popelis, I. Mazheika and E. Lukevics, *Chem. Heterocyclic Comp.* **35**, 436 (1999).
8. M. G. Dhara, D. Baskaran and S. Sivaram, *J. Polym. Sci., Part A: Polym. Chem.* **46**, 2132 (2008).
9. C. Moberg and I. lITkos, *React. Polym.* **16**, 171(1991/1992).
10. J. M. Wagner, C. J. McElhinny, Jr., A. H. Lewin and F. I. Carroll, *Tetrahedron: Asymmetry* **14**, 2119 (2003).
11. N. Dahlin, A. Bøgevig and H. Adolfsson, *Adv. Synth. Catal.* **346**, 1101 (2004).
12. I. Fallais, J. Devaux, R. Jérôme, *J. Polym.Sci., Part A: Polym. Chem.* **38**, 1618 (2000).
13. F. Moingeon, Y. R. Wu, L. Sanchez-Cadena and M. Gauthier, *J. Polym. Sci., Part A: Polym. Chem.* **50**, 1819 (2012).
14. L. Zhu, L. Liu and M. Jiang, *Macromol. Rapid Commun.* **17**, 37 (1996).
15. M. Ergin, B. Kiskan, B. Gacal and Y. Yagci, *Macromolecules* **40**, 4724 (2007).
16. M. Gauthier and M. Möller, *Macromolecules* **24**, 4548 (1991).
17. M. Gauthier, J. Li and J. Dockendorff, *Macromolecules* **36**, 2642 (2003).

31

Mater. Res. Soc. Symp. Proc. Vol. 1613 © 2014 Materials Research Society
DOI: 10.1557/opl.2014.155

Novel Hyperbranched Molecules Containing Pyrrole Units from Diacetylene Compounds and their Electronic Properties

Lioudmila Fomina[1], Jorge Godínez Sánchez[1], José A. Olivares[2], Fabio L. CuppoSant'Anna[2], Luis E. Sansores[1] and Roberto Salcedo[1]

[1] Instituto de Investigaciones en Materiales, Universidad Nacional Autónoma de México, Circuito Exterior s/n. C.U. A. Postal 70-360. Delegación Coyoacán. C.P. 04510. México D.F. México

[2] Centro de Investigación en Polímeros, COMEX, Marcos Achar Lobaton No 2 Tepexpan, México, 55885

ABSTRACT

Novel hyperbranched molecules containing pyrrole units were obtained from *ortho-, meta-* and *para-*diaminodiphenyldiacetylenes, as AB_2 type monomers by one-step polymerization. Diacetylenic fragments reacted with terminal amino groups in the presence of copper chloride to give pyrrole units. Diaminodiphenyldiacetylene monomers have been synthesized from ethynilanilines in three steps. The novel monomers and hyperbranched molecules were characterized by NMR, IR and thermal analysis. Some conductivity proofs were also carried out and this behavior was assessed.

The electronic behavior of some of these molecules was studied by means of theoretical methods. DFT optimization processes were carried out for three structures derived from the generation growing. There are at least two conformational isomers of the structure (*meta-* and *para-*) which show conductivity properties, the *meta-*isomer shows semiconductor nature but this species is hard to modeling because the steric hindrances cause optimization problems and indeed the third generation species was not achieved. In other context, the *para-*isomer allows the calculation of three generations and shows clearly a tendency to narrow the energy gap between the frontier orbitals but besides the behavior of the HOMO-1 seems reinforce the conductivity phenomenon.

INTRODUCTION

In the last decades have been made several researches about the synthesis and characterization of many organic materials bearing highly conjugated structures. These materials exhibit semiconduction and non-linear optics properties, so that these compounds are used to elaborate electrochemical devices, organic light emitting diodes (OLEDs), rechargeable batteries, smart windows, among others. The dendrimeric polymers can be classified into three large groups [1]: a). The first one refers specifically to the basic dendrimers i.e. molecules which grow in branches [2] and they are prepared by in steps; b). The second group is formed by the so called dendro-injerted polymers [3] which have been described as the intermediated species between the dendrimers and the hyperbranched polymers; c). And the third group is constituted by the hyperbranched polymers [4, 5].There are several species which belong to the third group [6]. They are very interesting substances with highly branched structures, third dimensional architecture, globular shape, terminal functional groups and particular electronic behavior [5]; they can be prepared by polymeric one or several steps synthesis [6].

The diacetylene derivatives shown in this communication belong to the third group. This work belongs to a research in which our group has published some interesting results [7, 8]. The new compounds can take orientations *orto-*, *meta-* and *para-*. All of them present a large electronic delocalization which make them important targets for high conductivity, particularly the *para*-analog has a curious geometry that can give it a large way for electronic transport, the theoretical study helps to explain the nature of this motion.

EXPERIMENTAL DETAILS

Reagents were provided by Aldrich Chemical Company and used as were received. FT-IR spectra were taken using a Nicolet 6700 spectrophotometer. NMR ^{1}H and ^{13}C spectra were recorded using a BrukerAvance 400 MHz spectrometer. The chemical shifts are reported in ppm on the scale relative to TMS. Melting points are uncorrected. Thermogravimetric analyses (TGA) were carried out in air at a heating rate of 20°C min^{-1} on a Mettler DTG 760 Instrument and Differential Scanning Calorimetry (DSC) was carried out at 20°C min^{-1} on a Mettler DSC 20 system.

Synthesis of the monomer compounds (figures 1-3)[6]
Synthesis of hyperbranched compounds(figure 4)[6]

Method **A**. A mixture of compound **3** (see figure 4) (0.1 g, 0.43 mmol), copper (I) chloride (0.05g, 0.5mmol) in 10 mL dimethylformamide was refluxed under nitrogen for 24-48 hours at 110°C in an oil bath and allowed to cool to room temperature. The mixture was diluted with excess of acidified water. The precipitate was collected by filtration and dried in vacuum. Method **B**. A mixture of compound **3** (0.5 g, 2.15 mmol), copper (I) chloride (0.05g, 0.5mmol) in dioxane (10 mL) was refluxed under nitrogen for 24 hours at 70°C in an oil bath and allowed to cool to room temperature. The solution was diluted with excess of acidified water. The precipitate was collected by filtration and dried in vacuum.

Computational Details

All structures have been optimized at B3PW91/3-21G using the Gaussian09 code. For all geometries the starting point was a non-symmetric structure. For the third generation, before the last optimization the structure was very near C2 symmetry, so we force to this symmetry and optimized again. The final structure is C2.

Impedance Measurements

For the impedance measurements it was used an LCR meter (QuadTech, 1910) that can operate in a broad range of frequencies. The measurement was made with 20Hz frequency, the cell containing the sample was placed inside the hot stage and heated at a ratio of 3.5 °C/min until reaching the temperature of 50°C then it was cooled at a ratio of 1 °C/min.

DISCUSSION

The synthetic route to the monomers 3 a, 3b and 3c and the hyperbranched molecules 4a, 4b and 4c are shown in figures 1-4. The monomers were prepared from *ortho*-, *meta*- and *para*-aminophenylacetylenes, in three steps [6].

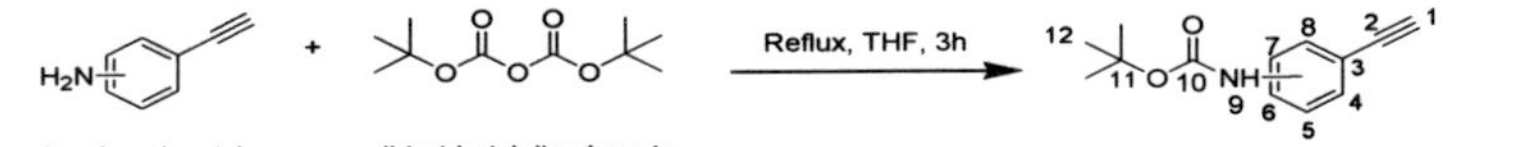

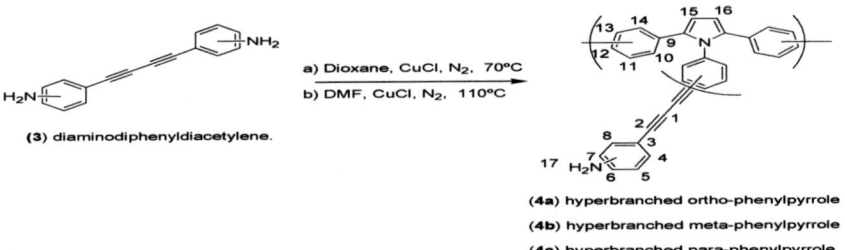

H$_2$N— (phenylacetylene)   +   di-tert-butyl dicarbonate   → Reflux, THF, 3h →

aminophenylacetylene   di-tert-butyl dicarbonate

(1a) 2-(N-Boc-amino)phenylacetylene.
(1b) 3-(N-Boc-amino)phenylacetylene
(1c) 4-(N-Boc-amino)phenylacetylene

Figure1. Synthesis of (*N*-Boc-amino)phenylacetylene (1).

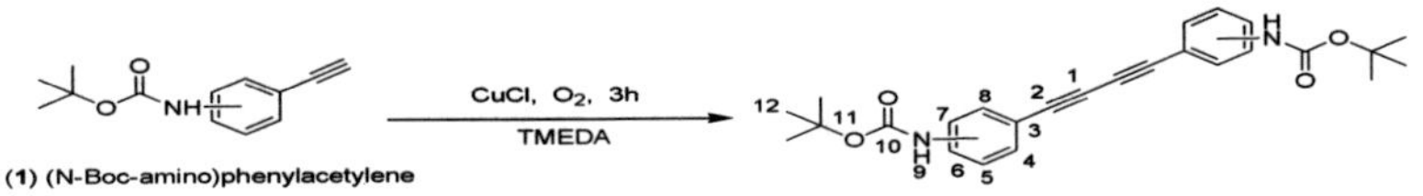

(1) (N-Boc-amino)phenylacetylene   → CuCl, O$_2$, 3h / TMEDA →

(2a) 2,2'-di(N-Boc-amino)diphenyldiacetylene
(2b) 3,3'-di(N-Boc-amino)diphenyldiacetylene
(2c) 4,4'-di(N-Boc-amino)diphenyldiacetylene

Figure 2. Synthesis of di(N-Boc-amino)diphenyldiacetylene (2).

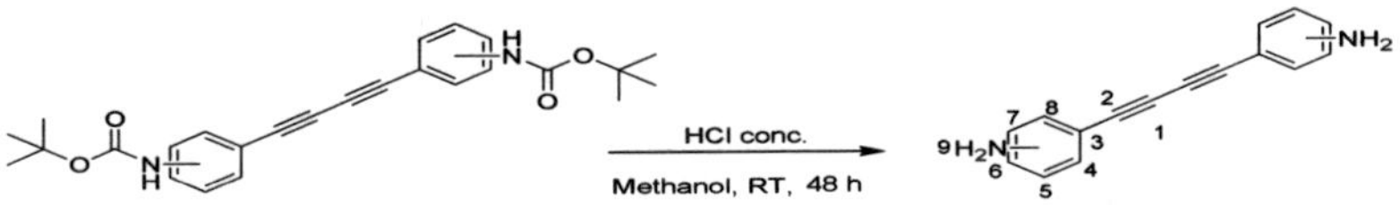

(2) di(N-Boc-amino)diphenyldiacetylene.   → HCl conc. / Methanol, RT, 48 h →

(3a) 2,2'-diaminodiphenyldiacetylene
(3b) 3,3'-diaminodiphenyldiacetylene
(3c) 4,4'-diaminodiphenyldiacetylene

Figure 3. Synthesis of diaminodiphenyldiacetylenes (3).

(3) diaminodiphenyldiacetylene.   → a) Dioxane, CuCl, N$_2$, 70°C / b) DMF, CuCl, N$_2$, 110°C →

(4a) hyperbranched ortho-phenylpyrrole
(4b) hyperbranched meta-phenylpyrrole
(4c) hyperbranched para-phenylpyrrole

Figure 4. Synthesis of hyperbranched molecules from diaminodiphenyldiacetylene (4).

Ortho-, *meta*- and *para*-diaminodiphenyldiacetylene (3a, 3b and 3c) as AB$_2$ type monomers were polymerized by two methods [6]. Method A - in DMF under nitrogen at 110 °C and method B - in dioxane under nitrogen at 70°C using copper (I) chloride as catalyst in both cases (figure 4). Diacetylenic fragments and terminal amino groups reacted to yield pyrrole units. These reactions are carried out by one step polymerization.

The conductivity of the hyperbranched compounds increases with the temperature in a giving range, that shows their semiconductor behavior. For *orto*-hyperbranched compound activation energy is 2.38 eV (for 45-100°C), for *meta*-hyperbranched compound 1.24 eV (*78-100°C*) and for *para*-hyperbranched compound 0.3 eV (8-34°C).

In figure 5 we show how the *para*-dendritic molecule (analog of *para*-hyperbranched compound 4c) is constructed, for a central bone fragments of dianyline-pyrrole are added.

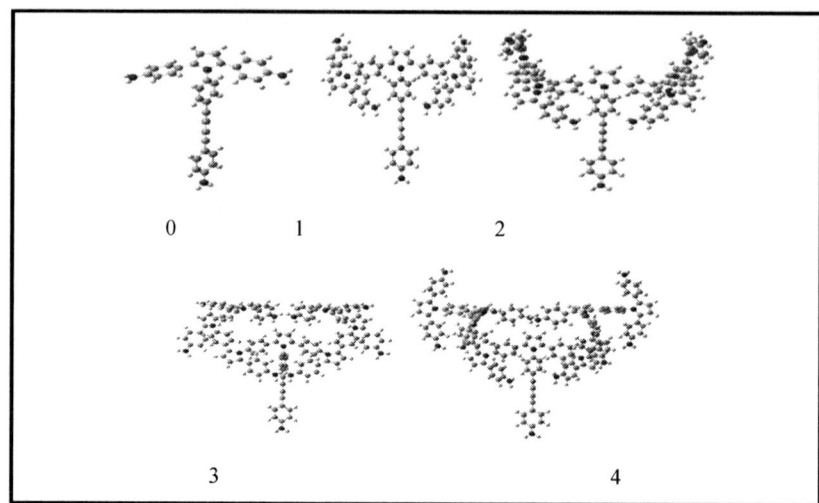

Figure 5. Optimized geometries of the generations 0 to 4 of the *para*-dendritic molecule.

For step 0 a dianyline-pyrrole fragment is added to the central bone. From there on, two dianyline-pyrrole fragments are added substituting two NH2, always selecting the external NH2 and keeping the symmetry to be a new nitrogen atom of a pyrrole ring. Generation 1, 2, 3 and 4 are shown in the same figure. As the molecule grows it shows a helical shape conserving the C2 symmetry for generation 2, 3 and 4. In all cases the steric hindrance was avoided. The molecule has a central backbone with a rod shape, while the fragments are like wings covering the backbone. As the molecule grows it takes a helical shape. It is expected that the molecules grow in this same way for higher generations. The main feature of this system is focused in the distribution of the frontier molecular orbitals. At generation 0 the LUMO is located at the central backbone with main contributions coming from the p orbitals of the carbons. It has b symmetry and remains the same in the next generations (see figure 6). The HOMO in generation 0 is located at the dianyline-pyrrole fragment with an a symmetry. It is important to note that the next three occupied orbitals are very near in energy but at this step they are not degenerated yet. In fact the HOMO -1 degenerates with the HOMO from generation 1 to 4. Furthermore, HOMO-2 and HOMO-3 degenerate from generation 2 to 4. These are accidental degeneration since the point group C2 does not have double irreducible representation. The energy values of the orbitals are given in table I.

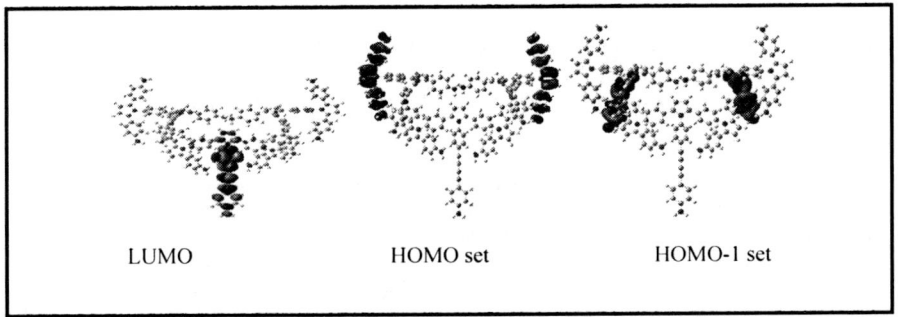

Figure 6. Spatial representation of the LUMO, HOMO set and HOMO-1 set for generation 4 of *para*-dendritic molecule.

Table I. The energy values of the orbitals of the *para*-dendritic molecule.

Generation	0	1	2	3	4
LUMO	-1.380	-1.523	-1.641	-1.657	-1.783
HOMO	-4.432	-4.340	-4.243	-4.270	-4.195
HOMO-1	-5.064	-4.341	-4.243	-4.298	-4.195
HOMO-2	-5.222	-5.239	-4.838	-4.533	-4.624
HOMO-3	-5.973	-5.256	-4.838	-4.533	-4.625
Gap	3.052	2.817	2.602	2.612	2.411
E Total	-39299.	-78602.	-117905.	-157208.	-196510.
Dipole	6.69	4.20	14.1	19.8	21.0

Figure 7 shows a scheme of the energy levels in generation 4.

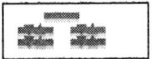

Figure 7. LUMO, HOMO set and HOMO-1 set energy sceme of *para*- dendritic molecule.

The important feature is that the molecule has 8 electrons available to participate in the conducting process, since the energy difference between the LUMO and the HOMO and HOMO-1 sets is very small (energy gap HOMO-LUMO is 2.41 eV, besides the HOMO and HOMO-1 gap is 0.43 eV). This phenomenon is the qualitative explanation of the experimentally measured high conductivity of the samples. The position of the orbitals is also important since the LUMO is always at the center of the molecule while the HOMO set is always at the external dianyline-pyrrole fragments. The HOMO -1 set is always located at the fragment of the previous generation. In this way there is an electron flows from the outside to the backbone.

CONCLUSIONS

Novel hyperbranched molecules containing pyrrole units were obtained from *ortho-*, *meta-* and *para-*diaminodiphenyldiacetylenes, as AB2 type monomers by one-step polymerization. Diacetylenic fragments reacted with terminal amino groups in the presence of copper chloride to give pyrrole units. Diaminodiphenyldiacetylene monomers have been synthesized from ethynilanilines in three steps. The novel monomers and hyperbranched molecules were characterized by NMR, IR and thermal analysis. The preliminary impedance spectroscopy study of electrical properties of obtained hyperbranched compounds shows clear evidence of semiconductor behavior of these compounds. The electronic behavior of these molecules was studied by means of theoretical methods. DFT optimization processes were carried out for three structures derived from the generation growing. The new compounds can take orientations *ortho-* ,*meta-* and *para-*. All of them present a large electronic delocalization which makes them important targets for high conductivity, particularly the *para-*analog has a curious geometry that can give it a large way for electronic transport.The *para-*isomer shows clearly a tendency to narrow the energy gap between the frontier orbitals but besides the behavior of the HOMO-1 seems reinforce the conductivity phenomenon.

ACKNOWLEDGMENTS

The authors wish to express their gratitude to Gerardo Cedillo for the assistance in NMR- and GPC- analysis and to Miguel Angel Canseco for thermal analysis.

REFERENCES

1. M. Seiler, *Fluid PhaseEquilibria* **241**, 155 (2006).
2. D. A. Tomalia and J. M. J. Frèchet, *Journal of Polymer Science Part A: Polymer Chemistry* **40**, 2719 (2002).
3. S.T. Teertstra and M. Gauthier, *Progress in Polymer Science* **29**, 277 (2004).
4. C.R. Yates and W. Hayes, *European Polymer Journal* **40**, 1257 (2004).
5. C. Gao and D. Yan, *Progress in Polymer Science* **29**, 183 (2004).
6. J. Godínez Sánchez, L. Fomina and L. Rumsh, *Polymer Bulletin* **64**, 8, 761-770 (2010).
7. L. Fomina, G. Zaragoza-Galán, M. Bizarro, Irineo P. Zaragoza, J. Godínez and R. Salcedo, *Materials Chemistry and Physics* **124**, 1, 257-263 (2010).
8. G.Huerta-Angeles, L. Fomina, L. Rumsch and M.G. Zolotukhin, *Polymer Bulletin* **57**, 433 (2006).

Mater. Res. Soc. Symp. Proc. Vol. 1613 © 2014 Materials Research Society
DOI: 10.1557/opl.2014.156

Copolymers based on 3-alkylthiophene and Thiophene Functionalized with Pyrene Chromophore

Edgar González-Juárez[1], Marisol Güizado-Rodríguez[1], Víctor Barba-López[2], Mario Rodríguez[3], Gabriel Ramos-Ortíz[3] and José Luis Maldonado[3]

[1] Centro de Investigación en Ingeniería y Ciencias Aplicadas (CIICAp), Universidad Autónoma del Estado de Morelos (UAEM). Av. Universidad No. 1001, Col. Chamilpa, C.P. 62209, Cuernavaca, Morelos, México.
[2] Centro de Investigaciones Químicas (CIQ), Universidad Autónoma del Estado de Morelos (UAEM). Av. Universidad No. 1001, Col. Chamilpa, C.P. 62209, Cuernavaca, Morelos, México.
[3] Centro de Investigaciones en Óptica A.C., Loma del Bosque No. 115, Col. Lomas del Campestre, C.P. 37150, León, Guanajuato, México.

ABSTRACT

As a preliminary study aiming to possible applications, novel polythiophenes (PTs) derivatives of 3-hexylthiophene and a thiophene functionalized with pyrene chromophore were synthesized. Homopolymer and copolymers of these monomers were obtained in different stoichiometric ratios which allow obtaining structure-property relation of each of the polymers. PTs were characterized by FT-IR, ^1H NMR, UV-vis, DSC-TGA, GPC and fluorescence experiments. Polymers have λ_{max} between 345 to 450 nm and an emission band at 485 and 542 nm. Low molecular weights distribution (Mn = 875 to 1600 g/mol) and thermostable products (T_d = 336 to 474°C) were obtained. These PTs functionalized with aromatic molecules and π-conjugated systems could offer interesting applications such as optical sensors, nonlinear optics and photovoltaic cells.

INTRODUCTION

Lately, conductive polymers (ICP) have been extensively studied due to their remarkable electrical, optical and biological properties; resulting in interesting applications as optical sensors, non-linear optics and voltaic cells [1]. Polythiophenes (PTs) are a known class of polymers that show good stability to the environment, good processability and easy functionalization in comparison with other conjugated polymers [2]. Additionally, the incorporation of select substituents into the polythiophene backbone improves some properties and induces chemical changes which can have important effects on the electronic properties for the polymeric material. Oxidative polymerization synthesis with ferric chloride is the most common method for obtaining PTs [3, 4]. In fact is the simplest, low cost and leads to the production of polymers and copolymers with high molecular weights [5], and generally good high regioregularidad polydispersity about 70% [6].

Recently, the synthesis, electrochemistry and fluorescence of thiophene derivatives decorated with coumarin, pyrene and naphthalene moieties have been reported [7]. Pyrene is a well-known fluorescent chemical compound containing several aromatic groups giving the ability to monitor optical processes even to microscopic level. The resultant photophysical properties can be used to known structural information about the conjugate system and give information about stacking interactions [8]. Nowadays, electrosynthesis of fluorescent copolymer based on pyrene and 2, 2´-bithiophene has been observed [9], nonetheless there isn´t information about chemical polymerization for polythiophene including the pyrene substituent.

In view of the above, herein we describe the chemical synthesis of homopolymer and copolymers derived of 3-hexylthiophene and a thiophene functionalized with pyrene chromophore. The obtained PTs were characterized by FT-IR, ^{1}H NMR, UV-vis, and Fluorescence. Thermal stability and molecular weight distribution of the polymers have been also investigated.

EXPERIMENTAL DETAILS

All reagents were purchased from Aldrich Chemical Co. and used without further purification unless otherwise mentioned. Solvents were purified by normal procedures and handled in a moisture-free atmosphere. ^{1}H NMR spectra of polymers were recorded at room temperature with a INOVA RMN spectrum of 200 MHz or a Varian Mercury 400 MHz spectrometer using 5–10% solutions in CDCl$_3$ and TMS as the internal standard. Absorption spectra were recorded on a spectrophotometer brand Instruments Spectrophotometer Perkin Elmer Lambda 900 UV-Vis, NIR double beam, double monochromator with two radiation sources. The Optical Experimental setup for fluorescence with a laser of 370 nm, 500 milliseconds integration time, scan average 1, Spectrometer Ocean Optics S2000. IR spectra (4000-400 cm^{-1}) were recorded on a Perkin Elmer Spectrum 400 spectrophotometer (ATR reflectance mode). For TGA analysis equipment was used SDT Q600 V8.2 Build 100, mode: Standard DSC-TGA with sapphire as calibration material and an alumina crucible. Nitrogen gas was used to 50 mL/ min, equilibrated at 50°C. The program used was Universal V4.2E. Average molecular weights and values of the polydispersity of polymers were determined by gel permeation chromatography (GPC) on a chromatograph Alliance 2695, using polystyrene standards for calibration and two columns PL gel linear, measurements were performed at 30 in THF and an injection volume of 25 µL.

Oxidative polymerization via iron trichloride procedure

In a three-necked flask is placed FeCl$_3$ dissolved in dry chloroform and nitromethane, allowing a continuous flow of nitrogen at 0 ° C. After, two addition funnels were placed in the three-necked flask and separate chloroform solutions of 2-(thiophen-3-yl) ethyl-4-(pyrene-2-yl) butanoate (obtained by an estrification reaction) and 3-hexylthiophene were added *via* cannula. Both solutions were allowed to pour dropwise to the solution of FeCl$_3$ by about 1 h at 0 °C. Subsequently, the polymerization reaction was stirred by 24 h at room temperature and was finished by addition of 200 mL of methanol standing the solution one day. A dark brown precipitate, doped polymer, was formed. After that, the solution was filtered and washed with methanol. Later, doped polymer was under stirring 1 h with 100 mL of a 1:1 solution of hydroxide ammonia/methanol. Finally, the undoped polymer was filtered, washed with methanol and again placed in a porous finger to a selective purification with chloroform using a soxhlet apparatus. The system was left under reflux 2 days, until no dissolving the copolymer was observed. The solutions obtained were concentrated using a rotary evaporator to yield the corresponding homopolymer or copolymers. Figure 1 shows synthetic scheme of copolymerization reaction. Table 1 summarizes the parameters used; copolymers poly(3-HT-*co*-

TPy) with different stoichiometric relations between monomers (3-hexylthiophene:thiophene functionalized with pyrene), 1:1, 2:1 and 5:1 and one homopolymer poly(TPy).

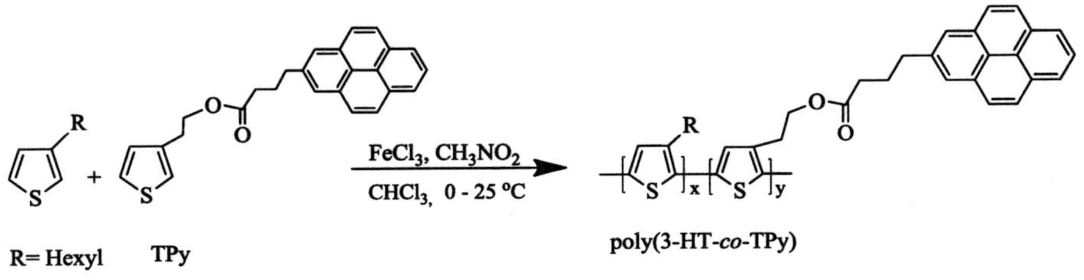

Figure 1. General scheme of copolymerization reaction.

Table I. Summary of the parameters for the synthesis of polymers.

Polymers	$FeCl_3$ mmol	Alkylthiophene mmol	Thiophene functionalized mmol	milligrams obtained
Poly (3-HT-co-TPy); 1:1	10.77	1.31	1.31	70
Poly (3-HT-co-TPy); 2:1	16.15	2.62	1.31	100
Poly (3-HT-co-TPy); 5:1	32.31	6.55	1.31	150
Homopolymer poly(TPy)	11.94	--------	2.61	150

DISCUSSION

The UV-vis absorption spectra for copolymers were measured in a chloroform solution (0.25 mg/mL) and are showed in figure 2a. In the copolymers poly(3-HT-*co*-TPy) relations 1:1, 2:1 and 5:1, the wavelength of maximum absorbance λ_{max} changes from 380 nm to 450 nm, while in the homopolymer its value is 345 nm; that absorption band can be assigned to electronic transitions π-π* in the thiophene ring. The difference between the intensity of absorption spectra could be due to the fact that the pyrene chromophore confers a higher absorption coefficient to the homopolymer and the copolymer relation 1:1. Moreover figure 2b shows the emission spectra of some polymers after irradiation with UV lamp of 370 nm. Homopolymer emits at 485 nm and the copolymers emit at 542 nm, figure 2b.

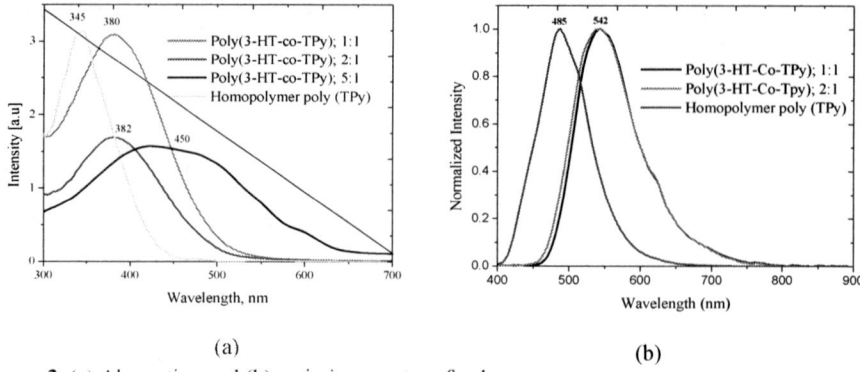

(a)   (b)

Figure 2. (a) Absorption and (b) emission spectra of polymers.

The figure 3 shows ^{1}H NMR spectra of the copolymer relation 2:1 and homopolymer which present broad signals that are characteristic of polymers. The incorporation of both monomer units can be observed along the spectrum. At 0.5-2 ppm chemical shift are observed corresponding to signals from the aliphatic chain of 3-hexylthiophene, the signals appearing at 2.5-4.8 ppm corresponding to the methylene protons of thiophenes functionalized or chain ester. Lastly, in the chemical shift of 6.9-7.5 ppm there are signals corresponding to the protons of the aromatic systems. With respect to the ^{1}H NMR spectrum of the homopolymer, it contains the same chemical shifts monomer unit. At 4.5 ppm is observed in a signal corresponding to the methylene protons found in the alpha position to the ester group, in the aromatic zone 6.9 to 8.2 ppm are observed aromatic protons of thiophene ring and pyrene chromophore.

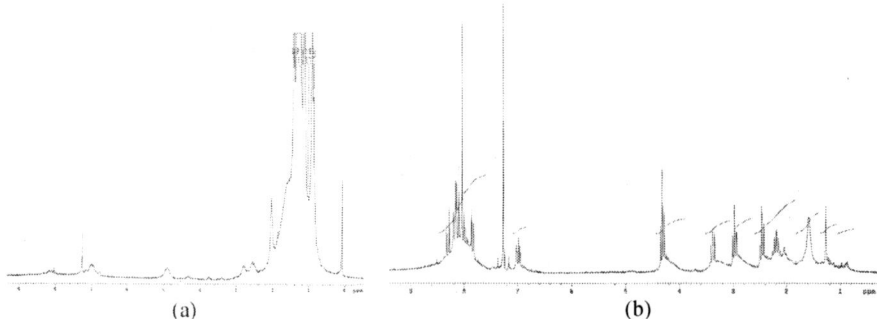

(a)   (b)

Figure 3. ^{1}H NMR spectrum of a) copolymer poly(3-HT-*co*-TPy) relation 2:1 and b) homopolymer poly(TPy), as an example.

The Figure 4 shows the FT-IR spectra of the polymers. Tension bands indicate the presence of functional groups. Table II summarizes bands of polymers.

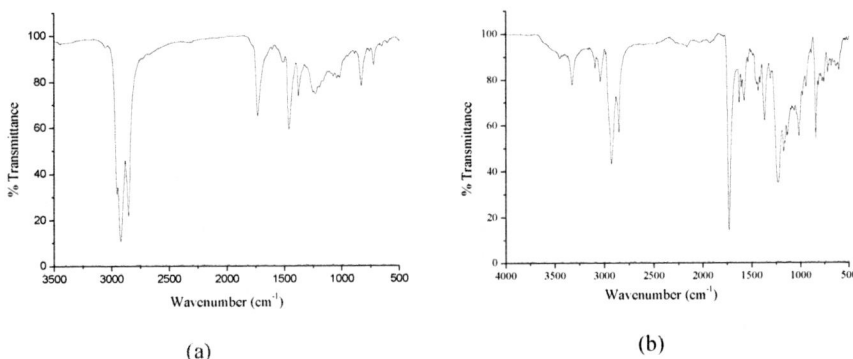

(a)   (b)

Figure 4. Spectra FT-IR of a) copolymer poly(3-HT-*co*-TPy) relation 2:1 and b) homopolymer poly(TPy).

Table II. Relevant FT-IR frequencies (cm^{-1}) for polymers.

Polymer	Vibrations bands						
	ν CH	ν =CH	ν C = C	ν C-S	ν C = O	ν C-O	ν C-O$_m$
Poly(3-HT-co-TPy); 1:1	2918	3056	1600	716	1728	1234	1021
Poly (3-HT-co-TPy); 2:1	2918	3449	1580	798	1738	1237	1099
Homopolymer Poly(TPy)	2933	3036	1634	836	1724	1245	1012

Thermal stability of synthesized polymers was determined by TGA- DSC analysis and molecular weights were determined by GPC analysis. The decomposition temperatures (T$_d$) and index polydispersity (M$_w$/M$_n$) are shown in Table III.

Table III. Data summary of TGA-DSC analysis and molecular weights.

Polymer	M_n g/mol	M_w/M_n	initial sample (mg)	T_d °C	Weight in loss	
					mg	%
Poly (3-HT-co-TPy); 1:1	1617	4.4	3.3	474	2.64	80
Poly (3-HT-co-TPy); 2:1	1666	3	5.97	426	2.89	48
Homopolymer Poly (TPy)	875	8.3	2.1	336	0.96	46

It was also found that the thermal stability of the copolymers is function of their molecular weights. The homopolymer has a molecular weight 875 g/mol and has a decomposition temperature of 336°C, which is much lower than that of copolymers poly(3-HT-co-TPy) 1:1, poly (3-HT-co-TPy) 2:1 having a molecular weight of 1617 and 1666 g/mol and a decomposition temperature of 474 and 426°C, respectively.

CONCLUSIONS

It was possible to carry out the oxidative polymerization with $FeCl_3$ as it is an easy, reproducible and inexpensive method to synthesize polythiophene. Through ^1H NMR it was possible to determine the integration of both monomer units. The UV-vis spectrum of polymers showed a band between $\lambda_{max} = 345-450$ nm, because of the different stoichiometric ratios. The FT-IR spectra showed bands of tension of each functional group of the polymer. TGA-DSC analysis showed that the copolymers are stable above 200°C and a decomposition temperature of about 400°C with a direct relation to the molecular weights.

ACKNOWLEDGMENTS

This project was funded by CONACYT CB2007-81383-Q.

REFERENCES

1. T. Yamamoto, *Macromol. Rapid Commun.* **23**, 583 (2002).
2. D.C. Casa, F.A. Morguera, C.P. Bizarri, M. Lanzi, *Synth. Met.* **124**, 467 (2001).
3. B. Wang, S. Watt, M. Hong, B. Domercq, R. Sun, B. Kippelen, D.M. Collard, *Macromolecules* **41**, 5156 (2008)
4. V.C. Gonçalves, L.M.M. Costa, M.R. Cardoso, C.R. Mendonca, D.T. Balogh, *J. Appl.*
a. *Polym. Sci.* **114**, 680 (2009).
5. A. Fraleoni-Morgera, C. Della-Casa, M. Lanzi, P. Costa-Bizzarri, *Macromolecules*
a. **36**, 8617 (2003).
6. G. Barbarella, A. Bongini, M. Zambianchi, *Macromolecules* **27**, 3039 (1994).
7. A.S. Abd-El-Aziz, S. Dalgakiran, I. Kucukkaya, B.D. Wagner, *Electrochim. Acta* **89**, 445 (2013).
8. F.M. Winnik, *Chem. Rev.* **93**, 587 (1993).
9. L. Xu, J. Zhao, R. Liu, H. Liu, J. Liu, H. Wang, *Electrochim. Acta* **55**, 8855 (2010).

Mater. Res. Soc. Symp. Proc. Vol. 1613 © 2014 Materials Research Society
DOI: 10.1557/opl.2014.157

Preparation and Characterization of Epoxy-Silica Coatings Using Rhodamine 6G as Dye

J.L. Varela Caselis[1], R. Agustín Serrano[1] and E. Rubio Rosas[1]
[1]Centro Universitario de Vinculación y Transferencia de Tecnología, BUAP Prolongación de la 24 sur y Av. San Claudio, sin número, Ciudad Universitaria, Col. San Manuel, Puebla, Puebla 72570, México.

ABSTRACT

It has been found that the hybrid materials are a compatible matrix for numerous organic compounds, such as organic dyes, laser dyes, and compounds that exhibit photo-chromic behavior and many more The epoxy-silica system seems to be an excellent matrix for organic dyes and a hybrid material suitable for to be used as coating on glass substrates with good adhesion properties. This work presents a systematic study of the effects of the different amount of using rhodamine 6G as dye on the structure and properties of epoxy–silica hybrids coatings synthesized by the sol-gel process. We have taken advantage on the high solubility of organic dyes in a hybrid organic–inorganic epoxy resin–silica (epoxy–SiO2) matrix to obtain homogeneous, hard and high optical quality red color films on glass substrates. The effects of the content of rhodamine 6 G on the optical and thermal properties of epoxy-silica hybrid films were also examined. Epoxy resin DER 332 cured with an amine (4,4 diamino diphenyl methane) was used as organic component and tetraethyl orthosilicate (TEOS) was used as precursor of the inorganic component. The results showed that at a concentration of rhodamine 0.05% coatings retain adhesion properties similar to coatings without colorant and the coatings are uniform and free of defects. These coatings have the potential to be used as filters and ornamental coatings.

INTRODUCTION

It has been found that the hybrid materials are a compatible matrix for numerous organic compounds, such as organic dyes, laser dyes, and compounds that exhibit photo-chromic behavior and many more [1-3]. Usually, inorganic–organic hybrid glasses are obtained in the form of thin coatings on different substrates by means of a low-temperature sol-gel process. Since the properties of the hybrids depend on the kind and amount of the units building their structure, the coatings are characterized by various properties, e.g. a refractive index changing within a wide range, anti-static and anti-reflectivity, corrosion protection, intensive color, and luminescence [4,5] For this reason the hybrid materials have found application as protective and decorative, colored coatings for glass items, as well as in new technical branches. Colored hybrid coatings are relatively low cost method for modifying the glass materials with a sophisticated shape and high surface area.

In this sense some researchers have conducted studies using dyes in organic-inorganic hybrid matrices. Almaral et al, [6]synthesized red colored transparent organic–inorganic hybrids films of a cross-linked polymethyl-methacrylate (PMMA) and silica via the sol–gel route using tetraethoxy-silane (TEOS) as precursor, a coupling agent and a commercial organic red dye. The results showed that the colored hybrid films consist of a homogeneous cross-linked organic–inorganic matrix with embedded dye molecules very well dispersed. The optical quality of the transparent colored hybrid films is very good with color intensity depending on the amount of organic dye in the films. The AFM measurements showed very flat and smooth film surfaces with rms average roughness about 0.3 nm. The color of the films is degraded when exposed to light illumination of certain intensity, however, for low light intensity this effect is small.

Wojtach et al [7], obtained colored inorganic– organic coatings on glass substrates, based on phenyltriethoxysilane, 3-glycidoxypropyltrimethoxysilane, and aluminium tri-sec-butylate, two groups of organic dyes were used, a commercial dye Orasol and no-commercial dyes obtained in the laboratory. They found that the intensity, as well as thermal stability of color and chemical resistance, depends on the kind of the introduced dyes.

Wojtach et al [8] obtained organic–inorganic hybrids by the sol–gel method from pure tetraethoxysilane (TEOS) modified with methyltrimethoxysilane or phenyltriethoxysilane. The hybrid materials were coloured by introducing organic absorption dyes and hybrid films were deposited on glass plates. The results indicated that the dyes were firmly incorporated in the hybrid matrix probably being trapped in some structural voids. The high intensity of colouring and stability of the hybrid films demonstrated the usefulness of the developed method of preparation and deposition in the manufacturing of coloured glass and ceramics.

The epoxy-silica system seems to be an excellent matrix for organic dyes and a hybrid material suitable for to be used as coating on glass substrates with good adhesion properties. However, to date, no reports have been found which use an epoxy resin-silica matrix as dye coating. Therefore a matrix based on commercial epoxy resin-silica system and 2-(3,4 epoxycyclohexyl)ethyl-trimethoxysilane seems to be particularly suitable for these applications.

In a previous paper [9] it was found that the epoxy-silica coatings to a concentration of 75%wt-25 %wt respectively, exhibited the best physicochemical properties as a protective coating on carbon steel substrates, based on this study, the composition of the hybrid coatings were the same (75% epoxy-25% silica).

We have taken advantage on the high solubility of organic dyes in a hybrid organic–inorganic epoxy resin–silica (epoxy–SiO2) matrix to obtain homogeneous, hard and high optical quality red color films on glass substrates. This work presents a systematic study of the effects of four different amount of rhodamine 6G (0.05, 0.1, 0.2 and 0.3 % wt) on the structure, adhesion, optical and thermal properties of epoxy–silica hybrids coatings synthesized by the sol-gel process and deposited on glass.

EXPERIMENT

The organic phase of the hybrid was prepared by curing epoxy resin DER 332 (Sigma-Aldrich) with 4,4' Diaminodiphenylmethano (DDM) from Fluka and 2-(3,4 epoxycyclohexyl)ethyl-trimethoxysilane (ECETMS) as coupling agent. The inorganic phase precursor of the hybrid was tetraethoxysilane (TEOS) from Sigma-Aldrich, with ethanol as solvent, hydrochloric acid as catalyst, glass slides (7 x 2.5) cm were used as substrates and rhodamine 6G was used as dye.

Synthesis of epoxy-silica hybrid

In a glass flask was prepared the mixture, TEOS:ECETMS:H2O:ETOH at molar rate 0.8:0.2:1:2 respectively, at room temperature, hydrochloric acid was added to until a pH of 2, the mixture was stir for 1 h (solution A). In other glass flask was prepared organic component DER 332-DDM (solution B). The hybrid was prepared mixing the solution A (75 %wt) with the solution B (25 %wt). Colored hybrid coatings were prepared adding to the hybrid solution rhodamine 6G. Different coatings were deposited on glass slides and were named as indicated in table 1.

Table 1. Code and formulation of the synthesized coatings

Code	Epoxy %wt	Silica %wt	Rhodamine 6G %wt
Epoxy	100	0	0
Silica	0	100	0
Hybrid	75	25	0
05R6G	75	25	0.05
1R6G	75	25	0.10
2R6G	75	25	0.20
3R6G	75	25	0.30

The dip coating method was used at rate of 22 cm/min and an immersion time of 2 min. The coated glass slides were pre-dried at room temperature for 1 h and cured at 100°C for 24 h. Scanning Electron Microscopy were performance using a JEOL, model JSM-6610LV, equipped with an Oxford X-ray microprobe. For the Fourier Transform Infrared Spectroscopy a BRUKER spectrometer model VERTEX 70 was used, operated in an interval of 500 to 4500 cm^{-1}. Optical transmission of coatings was measured within UV and visible ranges using a UV–VIS spectrophotometer PerkinElmer, Lambda 25 and Standard Test Methods Measuring Adhesion by Tape Test (ASTM D 3359) were used.

Measuring adhesion by tape test

Adhesion strength by tape test was used to evaluate the adhesion of the synthesized coatings, according to ASTM D 3359-02. The typical procedure is as follows: first, the surface of the sample-coated was cut by a razor to make grid lines. The total test area was 1 cm^2 with each square grid dimension of 1x1 mm. The Permacel 99 tape (a width of 25 mm, manufactured by Permacel) was applied firmly to cover the grid area of the coating at room temperature. After around 90 s, the Permacel tape was stripped off with one quick peeling. The adhesion strength of the coatings can be estimated by counting the number of squares peeled off as compared to the total number of squares.

DISCUSSION

Figure 1 shows SEM images of the surface of synthesized coatings. According to figures, the organic and inorganic particles of the all coatings were uniformly dispersed throughout the polymer matrix. The micrographs reveal that the coatings are uniform, and free of cracks. Figure 2, shows SEM micrographs of cross sections of the hybrid coatings, in this image can be observed that the hybrid, 05R6G and 1R6G are well bonded and are free of defects such as surface blistering or cracks, while the 2R6G shows detachment (figure 2d). The coating thickness of the hybrid and colored hybrid coatings measure on the glass slides was from 8 μm to 9 μm.

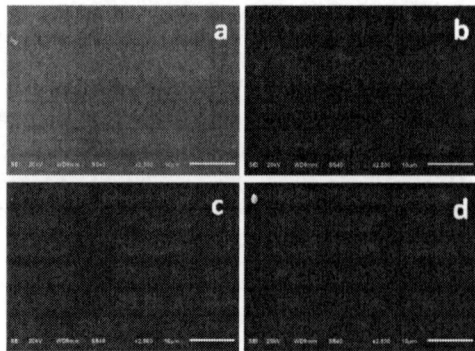

Figure 1. SEM images of the surface of synthesized coatings: a) epoxy; b) hybrid; c) 05R6G; d) 1R6G.

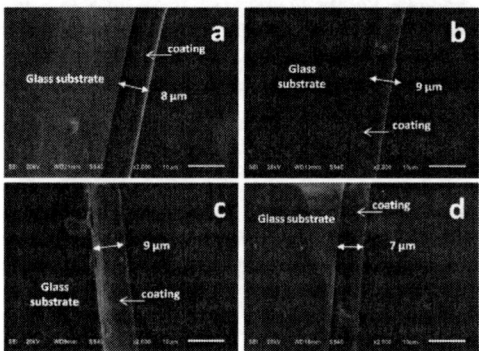

Figure 2. SEM images of transversal section of the coatings; a) hybrid; b) 05 R6G; c) 1 R6G; d) 2 R6G.

Figure 3 shows FTIR spectra for DER322, hybrid, 1R6G and silica. FTIR spectrum for DER 332 (figure 3a), shows characteristics bands of epoxy groups at 912 cm^{-1}. The FTIR spectra for hybrid and 1R6G coatings (figures 3b and 3c) indicated the disappearance of the band of epoxy group, instead, the absorption band of Si-O-C at 1081 cm^{-1} was observed, and these results indicated that covalent bonds were formed between inorganic and organic networks. The FTIR absorptions bands of the silica (figure 3d) show the characteristic peak of OH bond at 3230 cm^{-1} and 1625 cm^{-1}, the bands at 1100 cm^{-1} and 790 cm^{-1} are assigned to Si-O bonds and the Si-OH bond at 950 cm^{-1}.

Figure 4 shows thermogravimetric analysis for the different synthesized coatings, the test was from room temperature to 800 °C. The analysis shows that all coatings had highest weight loss above 400°C. The epoxy coating had a residual weight of approximately 15%, then this coating had the greatest degradation, the hybrid coatings 1R6G, 2R6G and 3R6G had a residual weight loss above 20%, while the coating 05R6G had a residual weight of approximately 36%, therefore this coating had the lowest degradation. This result shows that the thermal properties of the 05R6G coating were improved in relation to the epoxy coating. This improvement in thermal properties of the 05R6G hybrid coating was due to the addition of silica.

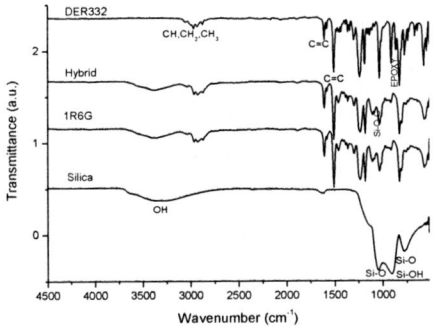

Figure 3. FTIR spectra for a) DER 332, b)hybrid, c) 1R6G and d)silica

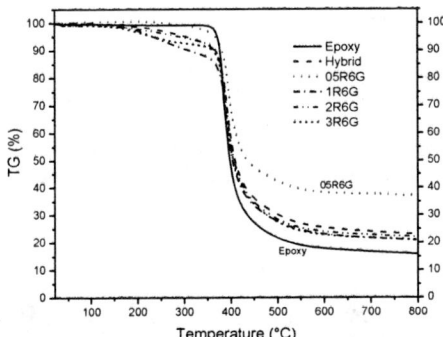

Figure 4. Thermogravimetric analysis for epoxy, hybrid, 05R6G, 1R6G, 2R6G and 3R6G.

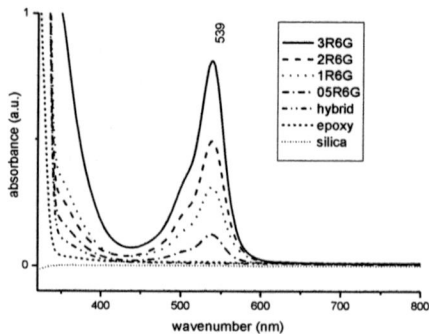

Figure 5. UV-Vis spectra of epoxy-silica hybrid coatings deposited on slide glass

The visible light absorption in the coatings is shown in figure 5, where it shows the UV-Vis spectra for the coatings synthesized. Hybrid, epoxy and silica coatings are transparent in this region of the spectrum, not so with coatings containing rhodamine 6G (3R6G, 2R6G 1R6G and 05R6G), these coatings have a maximum absorption at 539 nm in the visible region, likewise the absorption is directly proportional the rhodamine 6G concentration in the coating.

In this study, the measuring adhesion by test tape was used to examine the adhesion strength between the interface of the epoxy-silica hybrid coatings (including colored coatings) and the glass surface. It was found that the adhesion strength of the coatings (silica, hybrid and 05R6G) to the substrate surface was greater than epoxy system, as evidenced by the smaller stripped areas, as show in figure 6. However the adhesion of the 1R6G and 3R6G coatings was almost the same that the epoxy coating, this indicates that the amount of rhodamine 6G in the hybrid affects the adhesion strength of the coating to substrate. It is particularly noteworthy that the hybrid and 05R6G coatings showed a sufficient enhancement of adhesiveness (i.e. 10 and 5% of the area under the Permacel tape is peeled off vs 75 % epoxy coating). This percent was 45 % when the coating was 2R6G.

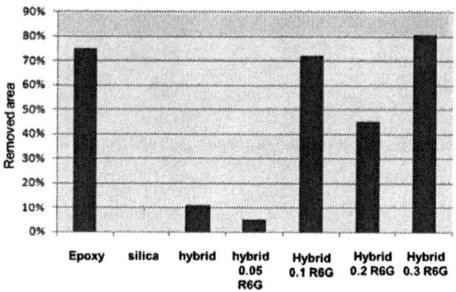

Figure 6. Adhesion test of the hybrid coatings according to ASTM D 3359-02

50

The 05R6G coating has adhesion properties similar to the coating without colorant (hybrid), this is in agreement with the SEM analysis, where Figure 2b shows that the coating is well adhered to the substrate therefore, these results indicate that the concentration of Rhodamine 6G at 0.05% does not affect the adhesion properties of the coating. Moreover, TGA analysis showed that this coating has the best thermal properties because had the greatest degradation temperature. The UV-Vis analysis shows that this coating provides adequate optical properties.

CONCLUSIONS

By the sol gel process was possible to synthesize epoxy-silica hybrid coatings uniform and free of defects on glass slides using rhodamine 6G as dye. This synthesis method is a simple process which is carry out at room temperature, these materials have the potential to be used as ornamental coatings on glass substrates. The concentration of Rhodamine 6G in the hybrid coating corresponding to 0.05 % had the best performance with the adhesion strength with a percentage of 5% of area removed.

REFERENCES

1. Y. Yang, M. Wang, G. Qian, Z. Wang, and X. Fan, Opt. Mater. 24, 621 (2004).
2. J. Kron, G. Schottner, and K. J. Deichmann, Glass Sci. Technol. 68C1, 378 (1995).
3. K. Taganaga, H. Yoshida, A. Matsuda, T. Minami, and M. Tatsumisago, Electrochem. Commun. 5, 644 (2003).
4. J. Kron, G. Schottner, K. and J. Deichmann, Thin Solid Films 392, 236 (2001).
5. K. H. Haas, S. A. Schwab, and K. Rose, Thin Solid Films 351, 198 (1999).
6. P. Hajji, L. David, J.F. Gerard, J.P. Pascault, and G. Vigier, J. Polym. Sci. Polym. Phys. 37, 3172 (1999).
7. L.M. Liu, Z. Qi, and X.G. Zhu, J. Appl. Polym. Sci. 71, 1133 (1999).
8. J.D. Cho, H.T. Ju, J.W. Hong, J. Polym. Sci. Polym. Chem. 43, 658 (2005).
9. J. Varela, V. R. Lugo, E. Rubio and V.M. Castano, Journal of New Materials for Electrochemical Systems, 14, 059 (2011).

Mater. Res. Soc. Symp. Proc. Vol. 1613 © 2014 Materials Research Society
DOI: 10.1557/opl.2014.158

Radiation-Grafting of Thermo- and pH-Sensitive Poly(N-Vinylcaprolactam-co-Acrylic Acid) onto Silicone Rubber and Polypropylene Films

Caroline C. Ferraz[1], Ademar B. Lugão[1], Gerardo Cedillo[2] and Emilio Bucio[3,]

[1]Instituto de Pesquisas Energeticas e Nucleares – IPEN/CNEN-SP, Av. Prof. Lineu Prestes, 2242, Cidade Universitaria, 05508-000, Sao Paulo, Brazil.
[2]Instituto de Investigaciones en Materiales, Universidad Nacional Autónoma de México, Circuito Exterior, Ciudad Universitaria, México DF 04510, México.
[3]Departamento de Química de Radiaciones y Radioquímica, Instituto de Ciencias Nucleares, Universidad Nacional Autónoma de México, Circuito Exterior, Ciudad Universitaria, 04510, México D.F., México.

ABSTRACT

This work focuses on the effect of gamma-ray radiation conditions on the stimuli-responsive of polypropylene (PP) films and silicone (SR) rubber substrates grafted with N-vinylcaprolactam (NVCL) and acrylic acid (AAc). PP films and SR rubber were weighed and placed into glass ampoules and exposed to ^{60}Co γ-source in the presence of air at room temperature, at dose rate around 12 kGy h^{-1} and dose between 5 and 70 kGy. Solutions of NVCL and AAc (1/1, v/v), 50 % monomer concentration (v/v) in toluene were added to the samples, the ampoules were degassed by repeated freeze-thaw cycles (5 times per 20 min) and sealed. The ampoules were heated at 60 or 70 °C at reaction time per 12 h. To extract the residual monomer and homopolymer formed during the grafting, the samples were soaked in ethanol for 24 h and then in distilled water, followed by drying under vacuum to constant weight. The values of grafting percentage achieved at a given irradiation dose were higher for SR than for PP. Samples where characterized by FTIR-ATR, DSC, swelling, LCST, and pH critical point.

INTRODUCTION

Radiation cross-linking, copolymerization or graft polymerization techniques have many advantages over other conventional methods, such as chemical and photochemical processes [1-4]. Furthermore, radiation induced graft polymerization is a convenient and powerful technique appropriate for modification of polymers. This methodology allows the introduction of specific chemical units for specific applications [5, 6]. ''Intelligent'' or smart polymers are stimuli-responsive materials that undergo volume changes in response to changes in temperature and pH. These unique characteristics are of great interest in drug delivery [7-9]. Thermo-responsive films exhibit changes with temperature variations and have the advantage of being easily manipulated to specific temperatures [10-12].

On the other hand, radiation grafting is a promising method for modification of materials, and it is of particular interest to achieve specifically desired chemical properties as well as excellent mechanical properties. The radiation grafting has been used for the preparation of smart materials, such as thermo- and pH-responsive materials, to be used as selective metal ion adsorbing materials and proton-conducting membranes [13]. The pre-irradiation technique is a clean and effective method for polypropylene (PP) and silicone (SR) modification, and number of reports on grafting polar monomer onto pre-irradiated films has been published [14]. Radiation modification of PP samples may be carried out in such a way that it acquires functional groups where a drug must be immobilized [15, 16]. Stimuli-responsive polymers are of particular interest due to their ability to undergo a well defined change in their properties [17]. Acrylic acid (AAc) is one of the most popular monomers that have been grafted onto different polymeric matrices and its polymer or copolymers with pH sensitive response and carboxylic groups from poly(acrylic acid) (PAAc) of pH sensitive response have a capability to undergo further chemical reaction to produce new functional groups [18].

Recent studies have incorporated different monomers such as N-vinylpyrrolidone (NVP) and amino acid derivatives as spacers between the carboxylic acid groups. This made the polymeric backbone more flexible and allowed greater access for acid/base reaction. Water-soluble polymers with lower critical solution temperature (LCST) have attracted a great deal of attention in recent years, due to their potential application in biomedicine and biotechnology. Among them, poly (N-vinylcaprolactam) (PNVCL) stands out based on the fact that it is not only nonionic, water-soluble, nontoxic and thermo-sensitive but also biocompatible. Moreover, the LCST of PNVCL is in the range of physiological temperature (32–34 °C) [19].

EXPERIMENTAL

Materials.

Polypropylene films and silicone rubber from Goodfellow England, 1 mm thickness and 1 cm x 5 cm in size were washed in methanol for 5 h and then dried in vacuum to constant weight. N-vinylcaprolactam and acrylic acid were from Aldrich Chemical Co., USA. Monomers were purified by vacuum distillation before use. Toluene and methanol from Baker were used as received.

Grafting.

PP films and SR rubber were weighed and placed into glass ampoules and exposed to ^{60}Co γ-source (Gammabeam 651 PT, MDS Nordion) in the presence of air at room temperature, at dose rate around 12 kGy h^{-1} and dose between 5 and 70 kGy. Solutions of NVCL and AAc (1/1, v/v), 50 % monomer concentration (v/v) in toluene were added to samples, the ampoules were degassed by repeated freeze-thaw cycles (5 times per 20 min) and sealed. The ampoules were heated at 60 or 70 °C at reaction time per 12 h. To extract the residual monomer and homopolymer formed during the grafting, the samples were soaked in ethanol for 24 h and then in distilled water, followed by drying under vacuum to constant weight. The grafting yield (Y_g) was calculated using the equation:

$$Y_g\ (\%) = 100[(W_g\text{-}W_o)\ /\ W_o], \qquad \text{Eq. (1)}$$

Where W_o and W_g represent the weights of the initial and grafted films, respectively.

Structure and thermal stability.

FTIR-ATR spectra were taken using a Perkin-Elmer Spectrum 100 spectrometer (Perkin Elmer Cetus Instruments, Norwalk, CT) with 16 scans. Differential scanning calorimetry (DSC) scans were recorded under nitrogen atmosphere using a DSC 2010 calorimeter (TA Instruments, USA) from 25 to 300 °C at 10 °C min^{-1}. LCST by DSC was measured between 23 and 60 °C at 1 °C min^{-1}

Interaction with water.

Water absorption equilibrium was monitored in duplicate by immersion of pristine and modified PP films and SR rubber into distilled water for 15 to 120 minutes. The excess of solution on the copolymer films was removed with filter paper, and the swollen samples were weighed. The swelling percent was determined by the equation:

$$\text{Swelling } (\%) = [(W_s - W_d)\ /\ W_d]100 \quad \text{Eq. (2)}$$

Where W_s and W_d represent the weights of the swollen and initial films respectively.

The LCST of the grafted films was determined by measuring changed equilibrium swelling of the samples immersed in distilled water at various temperatures from 27 to 35 °C for 2 h and neutral pH. The vials with films were set in a temperature controlled (25 °C) in phosphate buffer solutions of pH values from 2 to 8 during 2 h to obtain the pH critical point. The swelling percentage was determined gravimetrically by the equation 2. The thermosensitivity of the samples was defined and calculated as the ratio of the grafted samples swelling at 27 and 35 °C.

DISCUSSION

PP films and SR rubber were submitted to Gammabeam irradiation in air at different pre-irradiation dose prior to grafting. The films were then placed into toluene solutions of NVCL/AAc and reacted under various conditions. The grafting efficiency was assessed by gravimetry after extensive washing and drying of reacted films. It is expressed by the grafting yield Y_g, as defined in the experimental section (Eq 1).

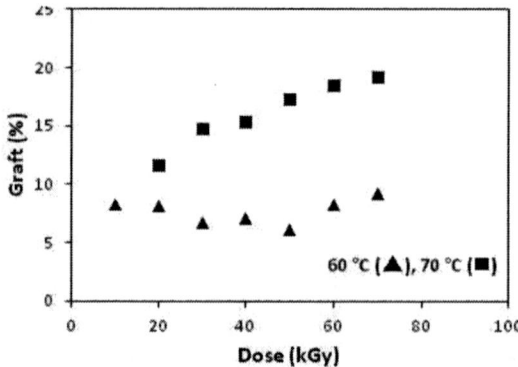

Figure 1. Evolution grafting as a function of poly(NVCL/AAc) onto pre-irradiated PP films, I= 12 kGy h^{-1}, t = 12 h, T = 60 and 70 °C.

The effect of pre-irradiation dose on the grafting yield was examined by performing graft polymerizations at 50 % monomer concentrations (NVCL/AAc, 1/1, v/v), for a reaction time of 12 h at a reaction temperature of 60 and 70 °C.

The relationship between NVCL/AAc grafting and the radiation dose was different for PP (Fig. 1) and SR (Fig. 2). For a fix NVCL/AAc concentration of 50% in toluene, the values of grafting percentage achieved at a given irradiation dose were higher for SR than for PP. For example, at 50 kGy (t=12 and T= 60 °C), the grafting percentages for PP-g-NVCL/AAc and SR-g- NVCL/AAc were ≈ 6 and 21 %, respectively.

FTIR-ATR analysis of the pristine and the modified PP and SR films were performed in order to confirm the presence of the NVCL/AAc grafted on both substrates (Fig. 3).

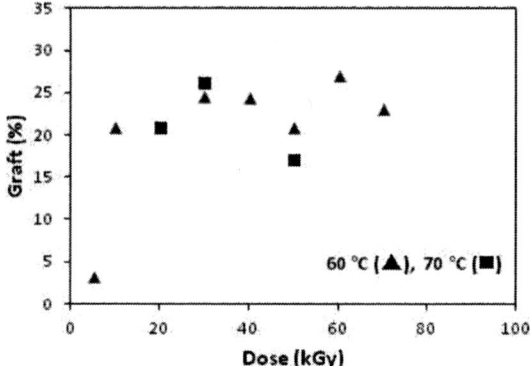

Figure 2. Evolution grafting as a function of poly(NVCL/AAc) onto pre-irradiated SR rubber, I= 12 kGy h^{-1}, t = 12 h, T = 60 and 70 °C.

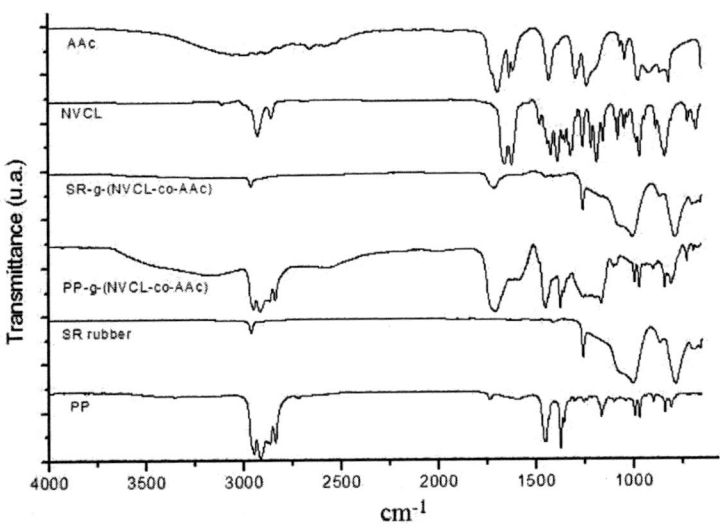

Figure 3. FTIR-ATR spectra of polypropylene film, silicone rubber, PP-g-NVCL/AAc (11 % graft), SR-g-NVCL/AAc (24 % graft), AAc, and NVCL.

The LCST behavior of the NVCL grafted onto PP films and SR rubber were determined by measuring the swelling ratio at temperatures ranging from 21 to 50 °C, at different grafted films of NVCL-co-AAc, there are no appreciable change in LCST value in films with different grafting percentages of both substrates in the range studied, critical temperature was 32 ± 1.5 °C.

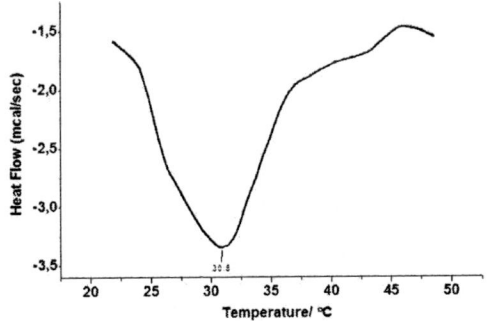

Figure 4. DSC thermogram of swelled PP-g-(NVCL/AAc) with distilled water at a heating rate of 1 °C min^{-1} from 21 to 50 °C.

The typical swelling behavior of AAc grafted onto PP and SR at several different pH values. It is obvious that the swelling ratios of PP-g-(NVCL/AAc) and SR-g-(NVCL/AAc) are significantly higher at a pH above 6 (after the pH critical point of 5.4), where the hydrogen bond interaction between two AAc units and between AAc and NVCL were destroyed, and the system present hydrophilic behavior, compared to the lower pH (before the pH critical point) with a hydrophobic behavior.

Figure 4 exhibits the DSC thermograms of the PP-g-(NVCL/AAc) 8 % graft, which was swelled in distilled water. The onset point of the endothermal peak, determined by the intersecting point of two tangent lines from the baseline and slope of the endothermal peak, was used to determine LCST [20]. PP-g-(NVCL/AAc) shows a LCST of 30.8 °C, the obtained LCST by DSC is almost equal to that which was obtained by swelling measurements.

CONCLUSION

Grafting of polymeric films (PP and SR) with sensitive polymers has been achieved by using a pre-irradiation method. The grafting of sensitive polymers onto polymeric films was confirmed by infrared analysis. The grafting efficiency increased with increasing the dose and temperature. Sensitive copolymers films presented pH- and thermo-sensitivity properties, which were determined by swelling and DSC respectively. Radiation grafting is an excellent method to obtain the binary grafting with the thermosensitive PNVCL and the pH sensitive PAAc. The properties of stimuli sensitivity were conserved after the grafting procedure.

ACKNOWLEDGMENTS

This work was supported by DGAPA-UNAM Grant IN202311 and CONACYT-CNPq Project 174378 (Mexico) and 490200/2011-7 (Brazil). The authors thank to M. Cruz, J. Rangel, F. García and B. Leal from ICN-UNAM for technical assistance.

REFERENCES

1. Meléndez-Ortiz H.I., Bucio E, Burillo G., *Radiat. Phys. Chem.*, 78, 1 (2009).
2. Huiliang W., Wenxiu C., *Radiat. Phys. Chem.*, 75, 138 (2006).
3. Palacios O., Aliev R., Burillo G., *Polym. Bull.*, 51, 191 (2003).
4. Bucio E., Burillo G., Adem E., Coqueret X., *Macromol. Mater. Eng.*, 290, 745 (2005).
5. Gupta B., Muzyyan N., Saxena S., et al., *Radiat. Phys. Chem.*, 77, 42 (2008).
6. Fujiwara K., *Nucl. Instrum. Methods B*, 265, 150 (2007).
7. Fu-Jian X., En-Tang K., Koon-Gee N., *Biomaterials*, 27, 2787 (2006).
8. Hoffman A.S., *Clin. Chem.*, 46, 1478 (2006).
9. Contreras-García A., Burillo G., Aliev R., Bucio E., *Radiat. Phys. Chem.*, 77, 936 (2008).
10. Chu L-Y., Park S-H., Yamaguchi T., Nakao S., *J. Membrane Sci.*, 192, 27-39 (2001).

11. Chu L-Y., Zhu J-H., Chen W-M., Niitsuma T., et al., *Chin. J. Chem. Eng.*, 11, 269 (2003).
12. Li P.F., Xie R., Jiang J-C., Meng T., et al., *J. Membrane Sci.*, 337, 310 (2009).
13. Jinhua Ch., Masaharu A., Yasunari M., Maseru Y., *J. Memb. Sci.*, 296, 77 (2007).
14. Chuanlun C., Qiang S., Lili L., Lianchao Z., Jinghua Y., *Radiat. Phys. Chem.*, 77, 370 (2008)
15. H. Wang, W. Chen, *Radiat. Phys. Chem.*, 75, 138 (2006).
16. Gupta B., Jain R., Anjum N., Singh H., *Radiat. Phys. Chem.*, 75, 161 (2006).
17. Ulbricht M. , Özdemir S., Geismann C., *Desalination*, 199, 150 (2006).
18. Adem E., Avalos-Borja M., Bucio E., Burillo G., Castillon F.F., Cota L., *Nucl. Instr. and Meth. B*, 234, 471 (2005).
19. Cheng S.C, Feng W, Pashikin I.I, Yuan L.H, Deng H.C, Zhou Y. *Radiat. Phys. Chem.*, 63, 517 (2002).
20. Zhang X., Ya Y., Chung T., Ma K., *Langmuir*, 17, 6094 (2001).

Mater. Res. Soc. Symp. Proc. Vol. 1613 © 2014 Materials Research Society
DOI: 10.1557/opl.2014.159

Self-Crosslinking Water Base Latex with Low VOC Emission

Juan J. Mendoza, Raquel Ledezma, Luis E. Elizalde

Centro de Investigación en Química Aplicada, Blvd. Enrique Reyna 140, CP 25253, Saltillo, Coahuila México. Tel. (844) 438 98 30, Fax: 438 98 39.

ABSTRACT

Here we report the preparation of functionalized latexes with isocyanate groups through an emulsion terpolymerization of BA/St/TMI. The emulsions were stabilized with two surfactants, EF-800 or MA-80; the TMI acted as a self crosslinked promoter during the film formation. We found a strong dependence of surfactant type in particle size and particle number. Moreover colloidal stability of latexes by a period of time was different, and the best colloidal stability was obtained with EF-800 surfactant. Swell index and mechanical properties of latex films were studied.

INTRODUCTION

Solvent less or low emission of volatile organic compounds (VOC) coatings is increasing attention due to a growing demand for environmentally friendly coatings. Additionally, increasing environmental pressures and related legislation push the technological or academic research to the latex-based coatings [1]. Latex coatings are promising alternatives to solvent borne coatings as they are inexpensive and have greater potential to lower VOC emissions [1,2]. However, in this kind of coatings performance is an important issue, and for solventless formulation the binder needs to be crosslinked during film formation. In the literature there are few reports of systematic studies of ambient cross-linking systems. Inaba et al. [3] obtain a crosslinked films of BA/St/TMI/MAA through miniemulsion polymerizations. They found, by drying the latexes at room temperature that the crosslinking reaction was greatly enhanced by the presence of a small proportion of MAA repeat units. Mohammed et al. [4,5] investigate the kinetics of polymerization and mechanical properties of MMA/BA/TMI polymers and their resulting crosslinked latex films when were immersed in triethylamine (TEA). Lovell and Yoon [6] report series of acrylic latexes based on poly (MMA/BA/TMI), prepared with different particle sizes, morphologies and TMI content. Apparently, the ambient self-curable latexes concept is controversial with the fact that the latex must be stable in aqueous media during the emulsion polymerization and the storage. Here we report the preparation of latexes for system BA-St-TMI through an emulsion terpolymerization. The TMI was used as a self crosslinked promoter during the film formation. The effect of the surfactant concentration, the monomer and initiator feed rate on polymerization kinetics and the storage stability of the latex, were investigated.

EXPERIMENTAL DETAILS

Butyl acrylate (BA) and Styrene (St) (both from Aldrich) were distilled under reduced pressure and stored at 4°C before to use. Dimethyl meta-isopropenyl benzyl isocyanate (TMI) and all other reagents were used as received without further purification. Surfactants, an anionic liquid sulfosuccinate mixture (Aerosol EF-800) and sodium dihexyl sulfosuccinate (Aerosol MA-80) were provided by CYTEC. Distilled-deionized (DDI) water was used in all experiments.

Polymerizations were carried out in a 500-mL jacketed glass reactor equipped with a reflux condenser and inlets for nitrogen, monomer and initiators feed, sampling and mechanical agitation. The polymerizations were carried out at 40°C in order to minimize the hydrolysis of the isocyanate groups of TMI. Either EF-800 or MA-80 was used as the surfactant. All of the surfactant was charged into the reactor along with most of the DDI water, and allowed to dissolve while being purged with nitrogen for a period of 30 minutes. After this time was added 5% of the monomer mixture, followed by addition of 33% of the initiators (Individual aqueous solutions of the potassium persulfate (oxidant) and potassium metabisulfite (reductant), were prepared with the remainder of the water charge). The rest of initiator solutions and monomer were dropwise over a different time period using KDScientific syringe pumps. Samples were withdrawn from the reactor at appropriate intervals to follow conversion gravimetrically and for measuring particle sizes.

Particle size (Dp) was measured by dynamic light scattering (Malvern S90) at 25°C. Number of particles (Np) was calculated from the polymer content and Dp. All latexes obtained were stored at room temperature and their colloidal stability was monitored by Dp measurements. Films were made within 24 h of latex preparation. The latex sample was placed on a mold and then was allowed to dry at room temperature. The polymer films were removed from molds 3 days after and dried an additional week at temperature and relative humidity controlled (23°C and 50%) until a homogeneous, transparent film was obtained.

Tensile measurements were generated according to the ASTM D638-03 method using a united universal tensile testing machine. A 24 lb load cell was employed, and the experiments were run at 50 mm/min, five specimens of each film were tested and the average values determined. The gel content of polymer films was determined gravimetrically; for this purpose, the film of known weight (Wi) was placed in a vial and then acetone was added and the vial was closed properly. It left undisturbed for 24 h. After that, the solution was centrifuged for 1 h at 20,000 rpm, and then was washed with fresh solvent to remove any soluble particle. Finally the solvent was extracted and the insoluble polymer was dried in air oven a 50°C for 1 h and weighed (Wf). The gel content was calculated using the following equation:

$$\text{Gel content (\%)} = (W_f/W_i) * 100 \qquad (1)$$

The gel content of each sample was determined at least three times and average value was taken.

DISCUSSION

Semicontinuous BA/St/TMI (ration 49/49/2) emulsion polymerizations were performed. In total, eight latexes were prepared with different surfactant contents and monomer feed. The variables to consider in latexes and results obtained for end conversion and Dp are summarized in Table I. The curves for variation of instantaneous conversion, particle diameter, and particle number with reaction time are shown for each of the eight semicontinuous reactions.

Table I. Variables to consider in polymerizations.

	Surfactant	Level of surfactant	Monomer feed rate	Conversion (%)	Dp (nm)
1	MA-80	2%	6h	95.4	96.4
2	MA-80	3%	6h	93.9	80.5
3	MA-80	2%	8h	91.9	94.4
4	EF-800	3%	6h	88.1	106.1
5	MA-80	3%	8h	92.9	74.3
6	EF-800	2%	8h	91.1	120.6
7	EF-800	2%	6h	85.7	117
8	EF-800	3%	8h	92.0	80

Conversion-time curves for these polymerizations are shown in Figure 1. Almost all conversions reach above 90% at end of reaction. As can be seen there is an increase in instantaneous conversion with increasing of surfactant concentration. Additionally, the addition time of monomers appears to be a factor which directly affects the conversion, with EF-800 is favored by a longer addition time, while in the case of MA-80 is observed favoring the conversion to 6 h, contrary to what happens with the EF 800.

Given that each of the semicontinuous emulsion polymerizations proceeded under monomer-starved conditions with controlled particle growth, the phases formed in their growth stages may be considered to be of reasonably uniform chemical composition. The instantaneous conversions are less than 100% due to mass transfer limitation [4].

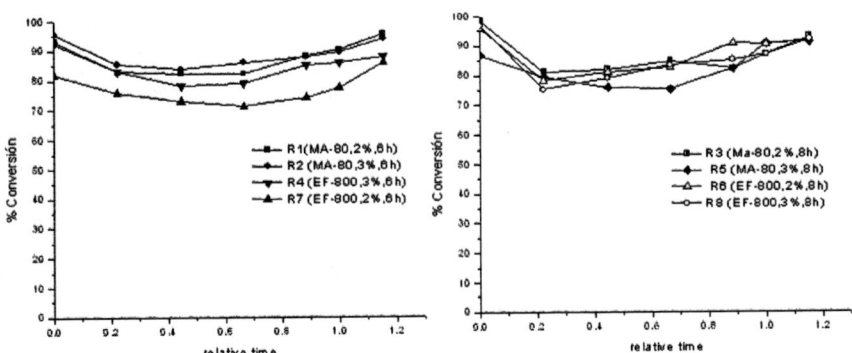

Figure 1. Variation of instantaneous conversion with time and with surfactant concentration.

The evolution of Dp as a function of conversion is showing in Figure 2. At the end of the polymerization, all latexes have Dp values between 70 and 120 nm. As expected, particle diameter increased in absence of secondary particle nucleation or particle coagulation. The differences in the particle diameter are a direct consequence of the differences in the type and concentration of the surfactant, the Dp decreased with the increment of the surfactant concentration (i.e. 2% to 3%). Dp larger are obtained with EF-800 than MA-80 as surfactant. The monomer feeding rate seems not to be a determining factor in Dp. In all experiments a narrow the particle size distribution was found.

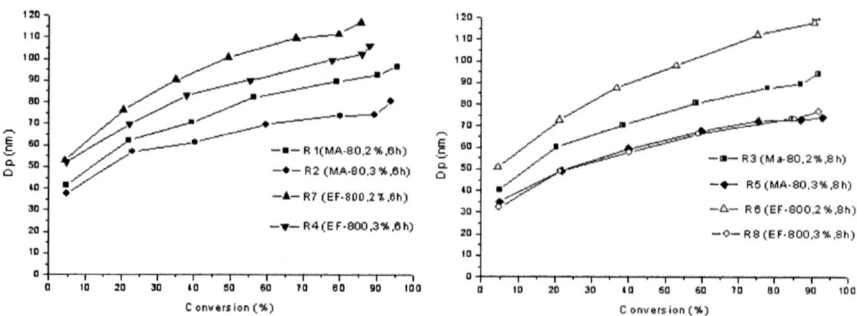

Figure 2. Particle size as a function of conversion for the polymerization of BA/St/TMI.

In order to examine the effect on the initiator feed rate on the particle nucleation period the particle number (Np) was estimated as a function of conversion (Figure 3). The Np was calculated from Dp considering spherical particles. The particle number was increased until approximately 20% of the conversion, after, the Np remained constant, and this indicates that particle nucleation ended at 20% conversion. Once the all particles were formed, they growth (increase of Dp) without the generation of new particles. However, in some cases, after 90 % conversion, depending on the type of surfactant, an increment of particle size Dp and reduction of Np was observed, this data suggests a coalescence process. The increment in the surfactant concentration drives to the increment of the number of particles, with a decrement in its size. The use of MA-80 surfactant generates a greater number of particles Np for both addition times.

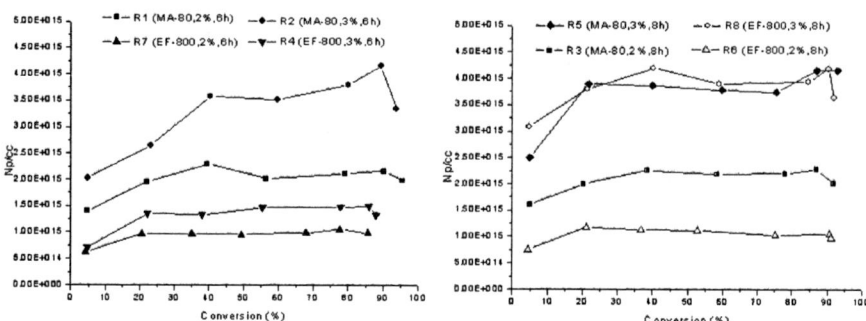

Figure 3. Number density of particles as a function of conversion.

For potential coating applications a colloidal stability for a long period of time is required. The latexes obtained in this work were relatively stable during polymerization and coagulum formation was not observed. Moreover, we study the stability of the latex over time, 5 to 6 months. Figure 4 shows the evolution of Dp/Dp$_0$ during storage at room temperature of latexes (where Dp$_0$ is the particle size at the end of the polymerization). It can be seen that the stabilized latex with MA-80 surfactant had problems with their colloidal stability, Dp values increased progressively to reach four to six times its original value after 5 and 6 months of storage. This type of instability also was observed by Treviño et al.[7] when they studied the copolymerization of vinyl acetate (VA) with butyl acrylate (BA) and TMI emulsions stabilized with sodium dodecyl sulfate (SDS) or sodium dodecyl benzene sulfonate (SDBS). In our case, the results obtained during the same period of storage with EF-800 surfactant indicates that, unlike what is observed with MA-80, particle diameter has been maintained with very little variation.

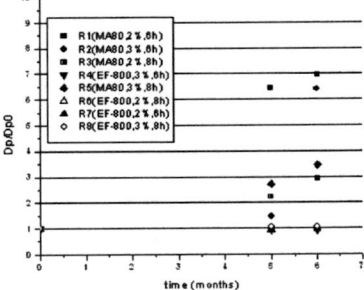

Reaction	Dp$_t$ (nm)	Dp (nm) 5th month	Dp (nm) 6th month
R1(MA80,2%,6h)	96.4	622.0	670.2
R2(MA80,3%,6h)	80.5	117.9	515.1
R3(MA80,2%,8h)	94.4	209.3	277.8
R4(EF-800,3%,6h)	106.1	97.8	95.0
R5(MA80,3%,8h)	74.3	257.2	202.5
R6(EF-800,2%,8h)	120.6	111.4	115.5
R7(EF-800,2%,6h)	117	111.9	116.0
R8(EF-800,3%,6h)	80	83.9	83.3

Figure 4. Latex colloidal stability during storage at room temperature.

Reaction R6 (Figure 4) was chosen for its high conversion, a larger particle size and better stability during storage. Poly(BA–St–TMI) and poly(BA–St) (control sample) were compared; the effects of TMI contribution on the physical and mechanical properties of latexes films were investigated. Formed films were flexible, transparent and free of imperfections that are usually observed when clots are present in the latex.

Gel content indicates the amount of crosslinked polymer. For these experiments we used acetone, because it was found able to completely dissolve a sample of copolymer formed solely by BA-St, without the addition of TMI. Gel contents (Table II) of the films were determined after the soluble polymer was separated. Gel fraction in each film increases with TMI contribution.

Table II. Gel content for poly (BA-St) and Poly (BA-St-TMI).

Relation	Reaction code	Gel content (%)
BA-St	R(EF-800, 2%, 8h)	0
BA-St-TMI	R6(EF-800, 2%, 8h)	33.83

The stress-strain curves are shown in figure 5. Tensile properties for samples are compared with the control sample without TMI. These films exhibit improved mechanical properties. When TMI is incorporated in system reaction, the tensile strength increased, however the ultimate strain at fracture decreased, as compared with the control sample. Therefore, a relative low concentration of TMI (used here) is enough to improving the mechanical properties. The average values of Young's modulus, maximum tension and maximum strain at break were also obtained by tensile measurements.

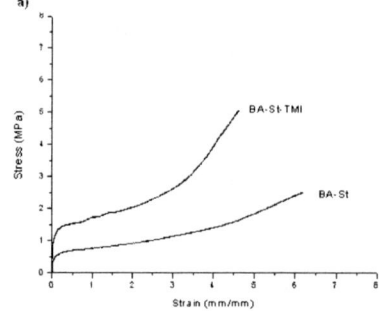

a)

b)

Latex	Young's modulus (MPa)	Tensile strength (MPa)	Ultimate elongation (%)
BA-St	18	2.2	593
BA-St-TMI	58	4.6	476

Figure 5. a) Stress-strain curves for films obtained from the latexes. b) Effect of TMI concentration on tensile properties.

CONCLUSIONS

Latexes were prepared successfully by semicontinuous emulsion polymerizations of BA-St-TMI using the potassium persulfate-potassium bisulfite redox couple for initiation and EF-800 or MA-80 as surfactant. It was found a strong dependence between the surfactant type in particle sizes and number particles. The latexes obtained with EF-800 surfactant were colloidal stable during storage for 6 months. Gel fraction of film appears with TMI contribution. The mechanical properties of films from the latex with TMI were better than sample control.

ACKNOWLEDGMENTS
The authors gratefully acknowledge the financial support of CONACYT. We also thank L.C.Q. José Luis Saucedo Morales for their assistance in tests tensile.

REFERENCES

1. I.P.A. Sørensen, S. Kiil, K. Dam-Johansen and C.E. Weinell, *J. Coat. Techol. Res.* **6(2)**, 135-176 (2009).
2. H.A. Mohamed, *J. Appl. Polym. Sci.* **125**, 170 (2012).
3. Y. Inaba, E.S. Daniels and M.S. El-Aasser, *J. Coating. Tech. Res.* **66**, 63 (1994).
4. S. Mohammed, E.S. Daniels, A. Klein and M.S. El-Aasser, *J. Appl. Polym. Sci.* **61**, 911 (1996).
5. S. Mohammed, E.S. Daniels, L.H. Sperling, A. Klein and M.S. El-Aasser, *J. Appl. Polym. Sci.* **66**, 1869 (1997).
6. P. Lovell and J. Yoon, *J. Macromol. Sci, Part B: Physics* **44**, 1041 (2005); **44**, 1065 (2005).
7. M.E. Treviño, J.C. Ramírez, H. Saade, R.G. López and L. Ríos, *Macromol. Symp.* **283–284**, 300-306 (2009).

Mater. Res. Soc. Symp. Proc. Vol. 1613 © 2014 Materials Research Society
DOI: 10.1557/opl.2014.160

Nanostructured LB Films Developed from Ferrocenylthioamide and Ferrocenylselenoamide Compounds

Rosa E. Lazo-Jiménez[1], María C. Ortega-Alfaro[1], José G. López- Cortés[2], José A. Chávez-Carvayar[3], Jordi Ignés-Mullol[4], Francesc Sagués[4], Violeta Álvarez-Venicio[1], María P. Carreón-Castro[1]

[1] Instituto de Ciencias Nucleares, Universidad Nacional Autónoma de México (UNAM), Circuito Exterior, C.P. 04510, Ciudad Universitaria. D.F., México.
[2] Instituto de Química, Universidad Nacional Autónoma de México (UNAM), Circuito Exterior, C.P. 04510, Ciudad Universitaria. D.F., México.
[3] Instituto de Investigaciones en Materiales, Universidad Nacional Autónoma de México (UNAM), Circuito Exterior, C.P. 04510, Ciudad Universitaria. D.F., México.
[4] Department of Chemistry-Physics, IN²UB, University of Barcelona, Martí Franqués 1, 08028. Barcelona, Spain.

ABSTRACT

In this work, the synthesis of two amphiphilic π-conjugated compounds such as ferrocenylthioamide and ferrocenylselenoamide, with the general formula $FcC=MNH(CH_2)_{15}CH_3$ with M = S or Se, are reported. The ferrocenyl group is a donor moiety forming a π-conjugated system with the amides of sulfur and selenium; both elements have also bioactivity with pharmacological interest. These two compounds formed Langmuir (L) monolayers at the air-water interface, which were characterized by isotherms of surface pressure versus molecular area (π-A) and compression/expansion cycles (hysteresis curves); Brewster angle microscopic images were also obtained. By using the Langmuir-Blodgett method molecular monolayers were transferred onto glass substrates. These nanostructures, in form of Langmuir-Blodgett (LB) films, were characterized through atomic force microscopy (AFM).

INTRODUCTION

Nowadays, research for new nanomaterials to create, design and synthesize new molecules -or modify them- to obtain thin films, based on organic and organometallic materials, have attracted great interest in many scientific and technological fields. Langmuir-Blodgett (LB) technique is an interesting alternative for the formation of new nanomaterials, like ultra-thin films. A major feature of this technique is the possibility to obtain well-ordered monomolecular films with controlled thickness [1]. In this paper, we report the synthesis of two amphiphilic organometallic compounds such as ferrocenylthioamide and ferrocenylselenoamide. We have added a long alkyl chain to the chemical structure of our compounds to be used with the LB technique, figure 1. Nanostructures from ferrocenylamide compounds (with M = S and Se) were developed as Langmuir films (L) at the air-water interface. Films were characterized by measuring the isotherm of surface pressure versus molecular area (π/A), which was recorded under similar conditions for both films and through BAM observations. Then, a monolayer was deposited on a substrate to obtain LB-films which were characterized by Atomic Force Microscopy (AFM).

In addition, ferrocene (Fc) containing compounds have been identified as an important π-conjugated donor group in the organometallic systems since they possess charge-transfer excited states [2]. The synthesis and applications of ferrocenyl-compounds have been developed and increased rapidly, because ferrocene is a good redox group known by its quasi-reversible oxidation and its good stability, not only chemical but under light [3]. Also, Fc derivatives have been applied in electrochemical and bio-redox systems [4]. Nevertheless, one of the most prolific areas of research in selenium chemistry is due to the potential biological activities observed in selenoderivatives, such as anticancer [5].

In a similar way, alkyl thioamides, new thiocarbonyl [6] and seleniumcarbonyl [7] moieties bonded to the ferrocene have been used as precursors of LB films. From ferrocenyl carbene complexes, previously synthesized [8], compounds were treated with NaBH$_4$-promoted demetalation of ferrocenyl aminocarbene complexes with elemental sulfur or selenium as an alternative to obtain ferrocenylthioamide or ferrocenylselenoamide compounds.

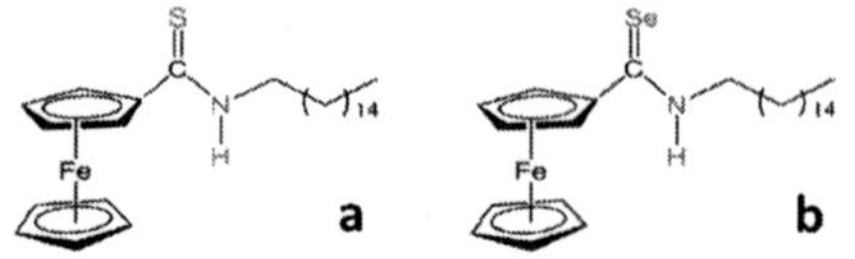

Figure 1. Molecular structure of: a) ferrocenylthioamide and b) ferrocenylselenoamide compounds.

EXPERIMENTAL DETAILS
Materials
Solvents and precursors used in this study were Aldrich products. Acetone was distilled over calcium chloride; tetrahydrofuran was distilled over sodium and benzophenone under a nitrogen atmosphere. Column chromatography was performed with Merck silica gel (70-230 mesh) and neutral alumina. To analyze their purity, all compounds were characterized by NMR spectroscopy. The ^{1}H and ^{13}C NMR spectra were recorded on a JEOL 300 spectrometer, using CDCl$_3$ as a solvent and TMS as an internal reference. Chemical shifts are presented in ppm (δ). IR spectra were performed on a Perkin-Elmer 283 B or 1420 spectrometer.

Synthesis of organometallic precursors
Ferrocenylhexadecylamine carbene complexes were prepared following a process described in a previous report [8]. The synthesis of aminoferrocenyl carbene was conducted through an aminolysis reaction with hexadecylamine, figure 2.

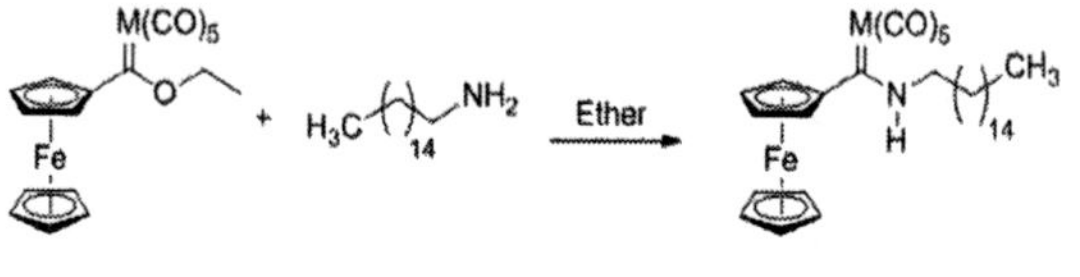

Figure 2. Synthesis of ferrocenyl hexadecylamino Fischer carbene complexe, M = Cr.

Then, different procedures were developed to remove the metal moiety and to transform Fisher carbene complexes into organic products. The synthesis of ferrocenylthioamide, figure 3, was obtained following a process previously described [6]. The new thioamide was characterized by conventional spectroscopic techniques.

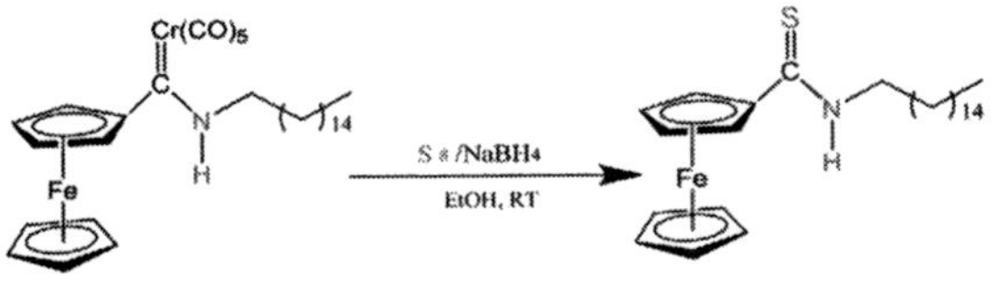

Figure 3. Oxidation of ferrocenyl aminocarbene by the mixture S_8/NaBH$_4$.

In general, the synthesis of ferrocenylselenoamide, figure 4, was achieved using the same protocol previuosly described [7]. Specifically, in this work we used hexadecylamine to add a 16-carbon aliphatic chain. An alternative way to synthesize ferrocenyl selenoamide was based on the oxidative demetalation of Fischer carbene complexes using elemental selenium/NaBH$_4$. Then, ferrocenylselenoamide was characterized by conventional spectroscopic techniques.

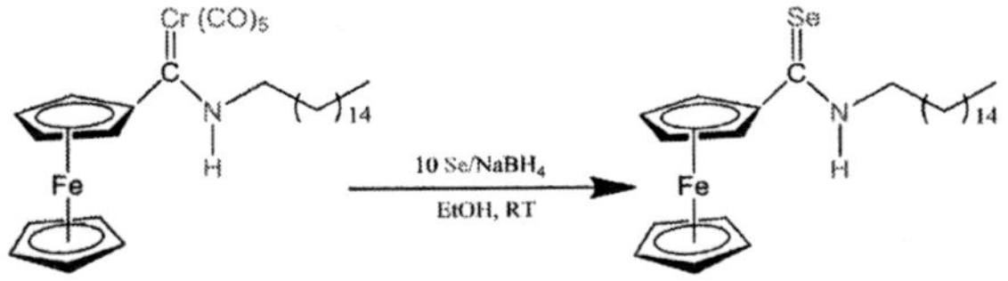

Figure 4. Oxidation of amino ferrocenyl carbene by the mixture 10 Se/NaBH$_4$.

Preparation of Langmuir films

The spreading solution was prepared in chloroform with a specific concentration of 1 mg mL^{-1}. ASTM type 1 ultra-pure water (Milli-Q system, 18.2 MW cm and simplicity 185, both from Millipore) was used for the subphase. Studies on the monolayer were carried out with a trough system, slowly spreading suitable amounts of solution on the water surface with a microsyringe (100 mL). After spreading, the monolayer was maintained for 10 min at room conditions for solvent evaporation. Afterwards, it was symmetrically compressed with a barrier speed of 5 mm min^{-1}. The surface pressure measurement, which was performed according to the Wilhelmy method, and the isotherm were recorded at 25°C. Finally, the stability of monolayers was studied through repetitive compression-expansion processes (hysteresis loops) without exceeding the collapse pressure.

The quality of the Langmuir films was monitored using a Brewster Angle Microscopy (BAM). The film quality observations were carried out at the Brewster angle (around 53.15° for an air-water interface) incidence [9]. During these experiments, high resolution images were directly acquired with a built-in CCD camera.

Preparation of LB films

The Langmuir monolayers obtained from ferrocenylthioamide and ferrocenylselenoamide were transferred onto solid substrates using the Langmuir-Blodgett technique with a vertical lifting method. Films were deposited at 25 °C, using 30 x 15 x 1mm^3 glass as substrates.

Characterization of LB films through AFM images of LB films, which were deposited on glass substrates, was acquired in tapping mode at room temperature and atmospheric pressure.

DISCUSSION
Langmuir monolayer of ferrocenythioamide

The isotherm, which was recorded at 25 °C, shows the preparation of the ferrocenylthioamide L-film, figure 5. This plot exhibited a gas phase behavior -at a surface pressure lower than 0.5 mN m^{-1}- followed by a transition to a liquid phase. When the surface pressure is increased, from approximately 4 mN m^{-1} to 8 mN m^{-1}, the first solid phase was observed. In this case, the molecular area extrapolated to zero surface pressure was approximately 56 ± 2 Å^2. Finally, it collapsed at a surface pressure of 11 mN m^{-1}. The hysteresis analysis showed that the Langmuir monolayer of the compound exhibited reversibility on successive compression/decompression cycles (hysteresis curve in figure 5, inset).

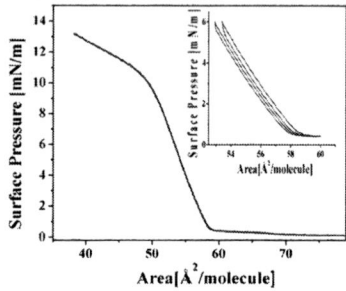

Figure 5. Surface pressure versus molecular area (isotherm) of ferrocenylthioamide. Inset shows the reversibility of successive compression-expansion cycles.

BAM images, figure 6, confirm the formation of a Langmuir film for the ferrocenylthioamide compound.

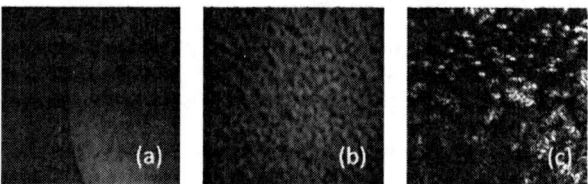

Figure 6. BAM images of ferrocenylthioamide film, at different surface pressures: a) 0.5 mN/m, b) 6 mN/m, c) 13 mN/m; (width screen: 100 μm).

Langmuir monolayer of ferrocenyselenoamide

The isotherm, which was recorded at 25 °C, shows the preparation of the ferrocenylselenoamide L-film, figure 7. It exhibited a gas phase behavior, at a surface pressure lower than 0.5 mN m^{-1}, followed by a transition to a liquid phase. When the surface pressure is increased, from approximately 4 mN m⁻¹ to 8 mN m⁻¹, the first solid phase was observed. The molecular area extrapolated to zero surface pressure was approximately 49 ± 2 Å². Finally, it collapsed at a surface pressure of 9 mN m⁻¹. The hysteresis analysis showed that the Langmuir monolayer of the compound exhibited reversibility on successive compression/decompression cycles (hysteresis curve in figure 7, inset).

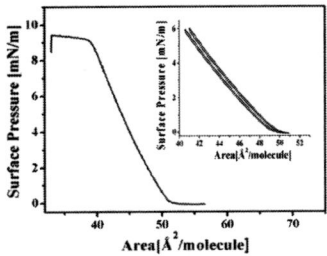

Figure 7. Surface pressure versus molecular area (isotherm) of ferrocenyseleoamide. Inset shows the reversibility of successive compression-expansion cycles.

BAM images, figure 8, confirm the formation of Langmuir film of the ferrocenylselenoamide compound.

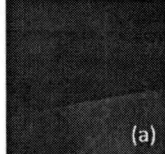 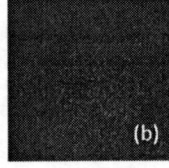 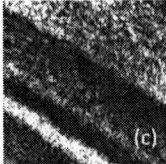

Figure 8. BAM images of ferrocenylselenoamide film, at different surface pressures: a) 0.5 mN/m, b) 6 mN/m, c) 9.5 mN/m; (width screen: 100 μm).

Langmuir-Blodgett films

It was observed that ferrocenylthioamide and ferrocenylselenoamide compounds formed Langmuir films in the condensed phase, which were transferred onto solid substrates through the LB method. AFM results depicted the molecular morphology for the condensed phase, figure 9a, and the collapsed phase, figures 9b and 9c, for the ferrocenylthioamide film. Although micrographs showed a homogeneous surface, after few days a low stability due to the possible formation of crystals on the monolayer was observed, figure 9c.

  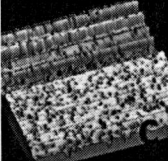

Figure 9. AFM images of ferrocenylthioamide monolayer: a) condensed phase, b) and c) collapsed phase.

In figure 10, AFM results depicted the molecular morphology of the ferrocenylselenoamide film, for the condensed phase, figure 10a, and the collapsed phase, figures 10b and 10c.

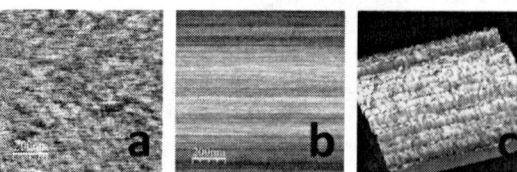

Figure 10. AFM images of ferrocenylselenoamide monolayer: a) condensed phase, b) and c) collapsed phase.

Variations in the morphology and roughness of the ferrocenylselenoamide LB film, studied through AFM in the collapsed phase, are in agreement with BAM results, where the main collapse mechanism was due to formation of multiple folds in ferrocenylselenoamide film, figures 10b and 10c, as described previously [10].

CONCLUSIONS

The synthesis of ferrocenylthioamide and ferrocenylselenoamide compounds were carried out and stable Langmuir monolayer at the air-water interface were observed and they exhibited reversible behavior upon successive compression-decompression cycles. After Langmuir-Blodgett ferrocenylthioamide and ferrocenylselenoamide films were obtained on hydrophilic substrates (glass and Si), AFM data confirmed the morphological results obtained through BAM, since both techniques provided evidence about the molecular arrangement of the films.

ACKNOWLEDGMENTS

We thank Martín Cruz (ICN-UNAM), Carlos Flores (IIM-UNAM), M. en C. Margarita Romero (F. Q.), Luis Velasco and Javier Pérez (IQ-UNAM) for their technical assistance. Authors thank DGAPA-PAPIIT IN111711 and IB200312 projects. R.E. Lazo-Jiménez thanks CONACYT for the scholarship 104243.

REFERENCES

1. P. Dynarowicz-Latka, A. Dhanabalan, O.N. Oliveira, *Adv. Colloid Interface* **91**, 221 (2001).
2. W. A. Amer *et al., J. Inorg. Organomet. Polym.* 20, 605 (2010).
3. S. Fery-Forgues and B. Delavaux-Nicot, *J. Photoch. Photobiol. A* **132**, 137 (2000).
4. K. Sakakibara *et al., Biomacromolecules* **8**, 1657 (2007).
5. Y.K. Lee *et al., Carcinogenesis* **31**, 1092-1099 (2010).
6. C. Sandoval-Sánchez *et al., J. Organomet. Chem.* **694**, 3692 (2009).
7. A. I. Gutiérrez-Hernández *et al., J. Med. Chem.* **2**, A-L (2012).
8. J.G. López-Cortés *et al., J. Organomet. Chem.* **690**, 2229 (2005).
9. J.J. Giner-Casares, G. Brezesinski, *Formatex Research Center,* Badajoz, Spain, 1007 (2012).
10. C. Ybert, W.Lu, G. Möller, C.M. Knobler, *J. Phys. Chem. B* **106**, 2004 (2002).

Biopolymers

Mater. Res. Soc. Symp. Proc. Vol. 1613 © 2014 Materials Research Society
DOI: 10.1557/opl.2014.161

Biopolymer Hydrogels Regenerated From Agave Tequilana Waste

For Cytocompatable Materials

Takaomi Kobayashi[1], Karla L. Tovar-Carrillo [1,2], Kazuki Nakasone[1], Motohiro Tagaya[1]
[1] Materials Science and Technology, Nagaoka University of Technology, 1603-1 Kamitomioka, Nagaoka, Niigata, Japan,
[2] Departamento de Ciencias Químico-Biológicas, Instituto de Ciencias Biomédicas, Universidad Autónoma de Ciudad Juárez, Ciudad Juarez 32300, Chihuahua,Mexico.

ABSTRACT

Agave fibers were used to elaborate a transparent and flexible cellulose hydrogel films used as scaffold for tissue regeneration and tested by in vitro assays with NIH 3T3 fibroblast cells. Using dimethylacetamide/lithium chloride (DMAc/LiCl) system was possible to obtain cellulose solutions and hydrogel films were prepared by phase inverse method without cross-linker. The concentration of LiCl in the DMAc solution was varied from 4 to 12 wt% in the phase inversion process and then the cytotoxicity was tested for 14 days on the cultivation. The resultant hydrogel films showed better cytocompatibility than the PS dish used as control. The cell growing images showed that the hydrogel films with lower LiCl apparently contained ordered and aggregated fiber orientation. This comparison suggested that the segmental microstructure in the hydrogel films influenced fibroblast cells spreading. In addition, the agave hydrogel films displayed good stability without biodegradiation through the cell cultivation.

INTRODUCTION

Hydrogels have been the primarily choice for a large number of researchers for many applications in regenerative medicine due to their unique biocompatibility [1,2]. Such attractive work shows that hydrogels can serve as scaffolds and provide structural integrity to tissue constructs [3-5]. In time, researchers have noted that synthetic polymers have offered limited properties for biological cell growing on the scaffold.Unfortunately this scaffold type still could not provide an ideal environment to support cell adhesion and tissue formation due to their bio-inert nature. In our strategy for biopolymer, cellulose having abundant hydroxyl groups, pure hydrogels of cellulose are tried to fascinating for the use of scaffold on fibroblast tissue engineering. Agave Tequilana waste fibers were used as biocellulosesource to elaborate hydrogel films [6,7]. Moreover, agave tequilana Weber azul, is an economically important plant cultivated in central Mexico for the production of tequila. More than 800 thousand tons of agave bagasses have been disposed as waste product per year [8]. This becomes as an important problem for the disposal of the bagasse agave and has been attention to regeneration to useful material. Within these waste products, bagasse from Agave tequilana Weber azul is found that until now, several research works have been performed to offer an alternative use for this waste product, while none is offered us a final solution, due to the necessity of the development of new technologies to solve this problem[9]. In our work [10], it was found that using dimethylacetamide/lithium chloride (DMAc/LiCl) system was possible to obtain flexible and transparent hydrogel films without chemical crosslinking. Several hydrogel films were prepared from the agave cellulose source by phase inversion method in ethanol as LiCl in the agave cellulose solution was varied. The resultant hydrogel films of agave cellulose presented that soft and good mechanical properties, event though there was no chemical crosslinking. For in vitro biocompatibility their results demonstrated that the hydrogel films tested showed good properties of biocompatibility and cytotoxicity especially in the lower LiCl content. In addition, the NIH3T3 fibroblast cells were used for cell adhesion assays.The growing cells showed better density and aspect ratio with cytocompatability. In the present work, comparison was made in the cytocompatability of the fibroblast cell that the cells grown on the hydrogel surface were significantly progressed even for two weeks cultivation, when the behavior was done with the commercial PS Dish used as control. These presented that the cellulose hydrogel films prepared from Agave Tequilana exhibited good cytocompatibility.

EXPERIMENTAL DETAILS

Tequilana Webber bagasse (Fig.1 (a)) was provided from Corralejo Tequila Company, in Penjamo, Guanajuato, Mexico. N, N-Dimethylethylendiamine (DMAc) was purchased from TCI, Tokyo, Japan, and stored for more than 5 days over potassium hydroxide before used. Lithium chloride was dried at 80°C for 12 h in a vacuum oven before uses.As shown in Figure 2, the agave fibers were treated by previous method in order to convert to agave cellulose fibers, which showed white color (Fig.1(b)) [10]. Then the fibers treated were used for phase inversion for following agave cellulose in DMAc/LiCl solution to solid hydrogels under ethanol vapor. The hydrogel film was immersed in distilled water over night and kept in phosphate buffered saline (PBS) at 4°C in a plastic container. Evaluation of fibroblast adhesion on agave hydrogel films was carried out as following process. The samples were sterilized with ethanol 50 and 70 vol% of concentration twice for 30 min, then rinsed twice with PBS 30 min, and finally swelled in DMEM for 2 h before the seeding procedure. NIH 3T3 mouse embryonic fibroblast cells were cultured at 37°C, 95% relative humidity and 5% CO_2 environment. The culture medium was 90% Dulbecco's modified Eagle's medium (DMEM) supplemented with 10% fetal bovine serum (FBS) and 1% penicillin/streptomycin. The cells were seeded on the agave films samples in polystyrene tissue culture dish (PS) 35 x 10 mm, at a density of $8x10^3$ cells cm^{-2}. PS dish was used as control. The cells were used for imaging and characterization purposes duringtwo weeks

Figure 1. Agave bagasse (a), agave cellulose and agave hydrogel film.

of the culture. The samples were imaged using inverse microscope. The images were analyzed for cell elongation and directionality using Cellsens software digital imaging software. To measure cell area the cell boundaries were marked. Aspect ratio, long axis and cell density were also measured. Approximately 50 cells were analyzed for each sample, and five images were analyzed to obtain an unbiased estimate of the cell density and morphology. The results presented herein were based on three independent experimental runs.To determine chemical changes of the polysaccharides solutions, FT-IR spectra were measured on a JASCO FT-IR/4100 spectrometerusing two CaF$_2$ plates with diameter 30 mm andthickness 2 mm (Pier Optics Co. Ltd.). In the FTIRspectra analysis for the OH stretching band, the fixed positionswere determined using Fourier self deconvolution for curve fitting [11-13]. These spectra on the wetting films were compared withthose of the dry films.

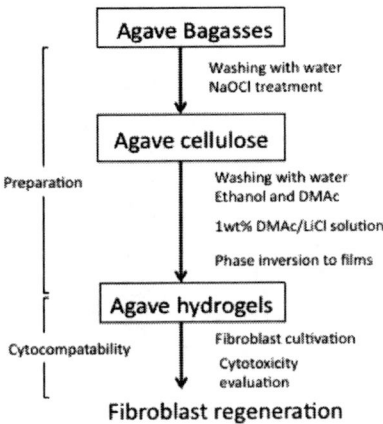

Figure 2. Preparation of agave hydrogels and evaluation of cytocompatability on the hydrogel scaffold.

DISCUSSION

In order to determinate cytotoxicity of the agave hydrogel films NIH 3T3 fibroblasts were used for the cell cultivation on the surface. Since fibroblast cell is a type of cell that can synthesize the extracellular matrix for animal tissue and the most common cells of conective tissue in animal, the investigation of the cytotoxic nature of the agave hydrogels was meaningful for tissue regeneration [14,15]. Figure 3 presents the cell density of the fibroblast cells on the agave hydrogel films prepared from the agave fibers varying LiCl concentration. In all cases, the results obtained in the hydrogel films were higher than the observed on PS dish used as control surface. With increasing duration time tile two weeks, the cell density became gradually to increase. As two weeks passed, the amounts of the cell density in the 4 wt% LiCl were actually higher. As indicated by Tamada and Salem, fibroblast was found to have a maximum adhesion on the hydrophilic surface [14,15]. Our previous report indicated that the film softness varied and changed to be soft nature at lower LiCl concentration [10]. This meant that the fibroblast cells had tendency to prefer soft surface for proliferation.

In Figure 4, results of cell morphology on the hydrogels films revealed a remarkable difference on fibroblast morphology for the hydrogel films (d-f) and commercial PS dish (a-c). For example, in the hydrogel film for the 6 wt% LiCl, the fibroblast surely adhered at the hydrogel surface at 4 hours and grown on the hydrogel film during 14 days, as observed in the images. After 4 hours of cell culture, the images (e and f) showed longer axis shape of the frown cells as compared with those adhered on the PS Dish at the same cultivation times. Moreover, the boundaries of the adhered cells on the cellulose films seemed to be tightly adhered on the hydrogel surface, showing a diffuse shape. In addition, anisotropic shape was observed in the first hours of the cell culture, when the hydrogel film was used in the scaffold on the cell

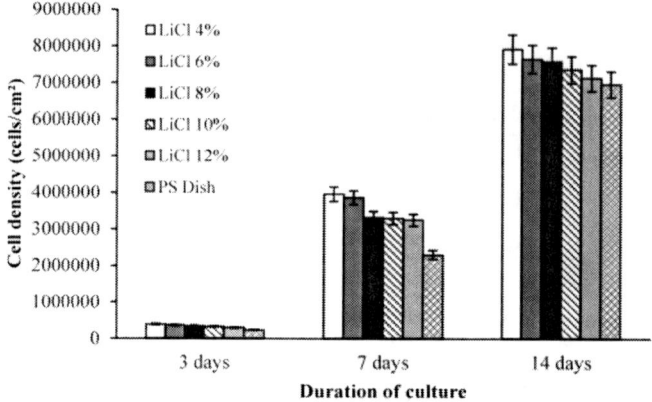

Figure 3. Fibroblast cell densities observed on the surface of agave hydrogel scaffold.

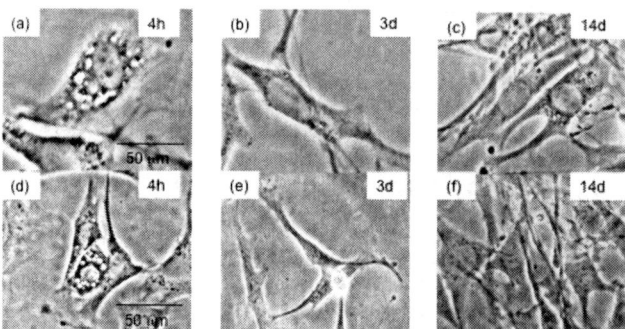

Figure4. Fibroblast cell area (a, d), aspect ratio (b, d) and long axis of growing cells for 3days (upper) and 14 days (bottom) cultivation.

growing. Relative to the hydrogel film, the fibroblast shape observed on the reference PS dish was mainly round at 4hours, as seen in (a). This indicated that the cell could not tightly adhere to the surface of the commercial dish. The cell growing through the cultivation period spresented higher cytocompatibility in the hydrogel film. As resulted in Figure 5, when the LiCl concentration was increased in the DMAc solution to 12 wt%, the resultant cell shape showed lower aspect ratio and long axis with lower cell density on the hydrogel films. This might be due to the decreased swelling and softness of the sample film. These results revealed that the hydrogel films made of the agave cellulose provided a better surface for fibroblast growing, depending upon hydrogel properties.This could be attributed to that at lower content of LiCl case the softness and stiffness of the films might regulate the adhesion between cellular- extracellular matrix molecules.

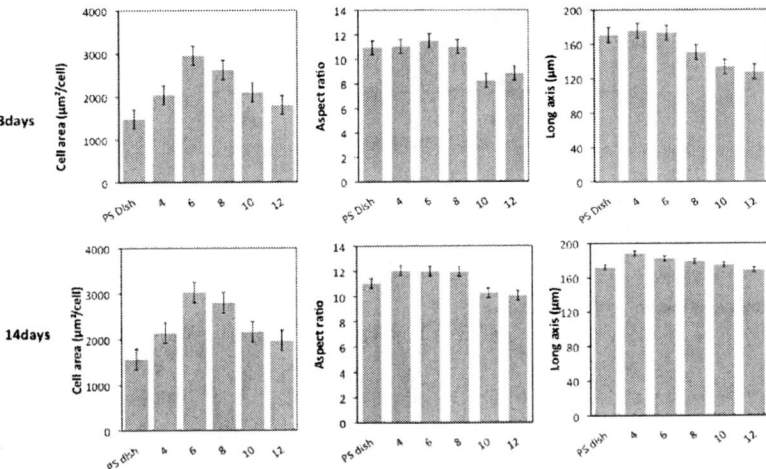

Figure 5. Cell area, aspect ratio and long axis of the cultivated fibroblast cells appeared on hydrogel films and PS dish at 3days (upper) and 14 days (bottom).

In tissue regeneration degradable biomaterials are desirable to replace by newly formed tissue upon regeneration. In addition, there is another interest to examine whether the hydrogel films showed highly or less biodegradation during the cultivation. Thus, the weight of hydrogel films was measured as shown in Figure 6. The hydrogel degradation was tested by weighing the film cultivated for 14 days. As seen, there were very less changes in the weight of the hydrogel films during the cell culture time for the fibroblast cells. As presented, the film weights were slightly increased in about 0.1-0.3% with increasing the cultivation time. This might be due to the fibroblast cell grown on the surface.

In order to make sure these data, we also measured FT-IR spectra of the wet hydrogel films. So the initial spectrum was compared with that cultivated for 14 days. Figure 7 exhibits few change in the spectra between the initial and the cultivation spectra. In the comparison

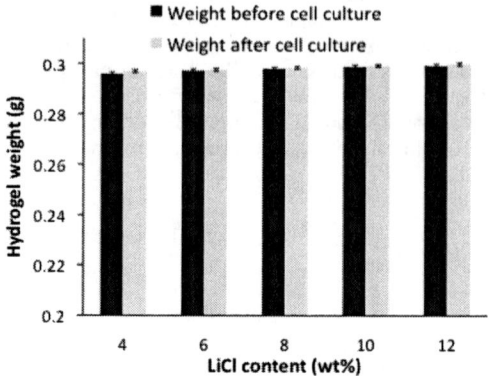

Figure 6. Comparison of hydrogel films in the weight at initial and 14 days cultivation for each film prepared with different contents of LiCl.

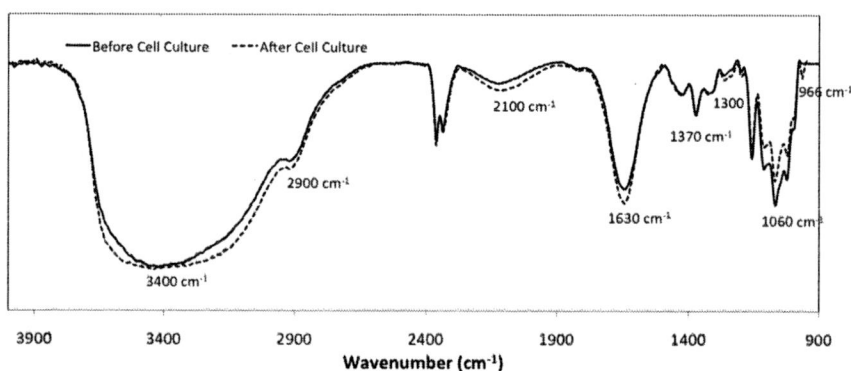

Figure 7. FT-IR spectra of agave hydrogel films before and after the fibroblast cell cultivation for 14 days.

before and after the cultivation of the fibroblast cells for 14 days, it was noted that strong peaks for water at 3400, 2100 and 1630 cm^{-1}were observed in the wet hydrogel films.The other IR peaks appeared around 2900, 1370 and 1060 cm^{-1} were originated with the agave cellulose for CH stretching, C-O stretching and C-O-C groups attributed to the glycosidic bonds [16], respectively. It was noted that the cultivated film had smaller peak at 966 cm^{-1} for the fibroblast cells on the surface. This was only difference in the both spectra. Therefore, this comparison implied that no degradation of the cellulose scaffold during the cell culture was observed. Furthermore, the figure in the FT-IR spectra exhibited the interaction change in water and OH group of the hydrogel film. As seen, the broadening in 3000-3800 cm⁻¹ was observed in the cultivated hydrogel film, indicating increment of the polymeric hydrogen bonding in the cellulose film cultivated.

As known, FT-IR spectrometry can provide useful information of polymeric hydrogen bond network [13]. The FT-IR spectra showed the distinctive bands appeared in the region of 3000–3800 cm⁻¹ for the OH stretching bands [13]. Figure 8 shows comparison of the peak deconvolution for the peak 1 (2900 cm⁻¹), peak 2 (3045 cm⁻¹), peak 3 (3300 cm⁻¹), peak 4 (3440 cm⁻¹), peak 5 (3550 cm⁻¹) and peak 6 (3600 cm⁻¹) for CH stretching, hydrogen bonds of inter-polymer and intra-polymer interaction in the cellulose, free OH group of the cellulose, water-cellulose hydrogen bond and water hydrogen bond, respectively [11,12]. In the broadening stretching band for the cultivated film, the result of the deconvolution indicated that the peaks 2 and 5 become somewhat intense in the cultivated film for the inter-polymer hydrogen bond and water-cellulose hydrogen bond having absorption at 3045 cm⁻¹ and 3550 cm⁻¹ [11-13]. Therefore, by the cultivation at 38°C for 14 days, it seemed that the internal hydrogen bonding of the cellulose segments could be enhanced in addition with water sorption on the OH groups of the segments.

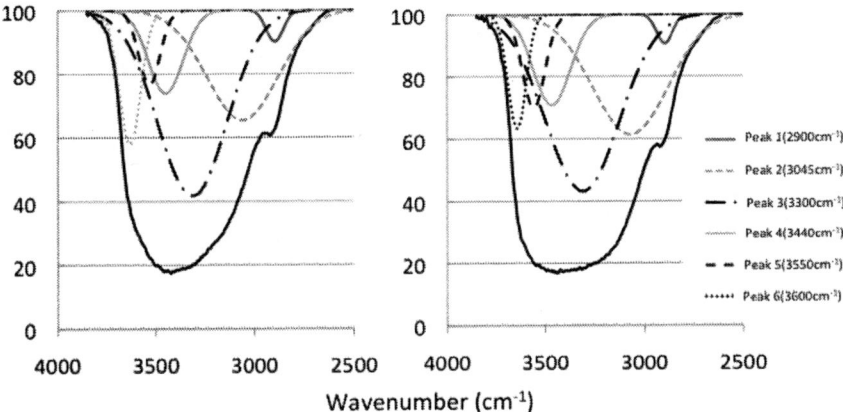

Figure 8. FT-IR spectra of the agave hydrogel films before (left) and after the cultivation (right) for 14 days. The LiCl concentration was 6wt% in the phase inversion process of the DMAc solution.

CONCLUSIONS

Hydrogel films prepared from agave fibers were obtained by phase inversion of the DMAc solution with LiCl. The cellulose hydrogel film having better mechanical properties could be prepared with no chemical crosslinking. Depending on the LiCl concentration, these hydrogel films exhibited softness on the viscoelacity [10]. Fibroblast cell was cultivated on their surface for 14 days at 38 °C, showing very good cytocompatibility in the hydrogel films relative to reference PS dish. It was proved that the hydrogel softness influenced initial cultivation of the cells. Then, since the LiCl acted to be dense networks of the agave cellulose segments in the hydrogels at 4 or 6 wt% concentration, the cell cultivation showing better cytocompatability was obtained in vitro test. These results indicated that the hydrogel films might become leading biomaterials as candidate in tissue engineering.

ACKNOWLEDGMENTS

This work was partially supported with Tokyo Ohka Foundation for the Promotion of Science and Technology. One of the authors (Karla L. Tovar-Carrillo) was supported with scholarship of the Consejo Nacional de Ciencia y Tecnología (CONACyT) by Mexican government.

REFERENCES

1. R. Langer, J. P. Vacanti, *Science* **260**, 920 (1993).
2. A. Sionkowske, *Prog. Polym. Sci.* **36**, 1254 (2011).
3. M. Lydon, T.Minett, *Biomaterial* **6**, 396 (1985).
4. L. De Bartolo, S. Morelli, A. Bader, E. Drioli, *Biomaterials* **23**, 2485 (2002).
5. B. H. Thomas, C. Fryman, K. Liu, J. Mason, *J. Mech. Behav. Biomed. Mater.* **2**, 588 (2009).
6. G. R. Filho, S.D. Ribeiro, C. da S.Meireles, L.G. Da Silva, R. Ruggiero, M. F. F. Junior, D. A. Cerqueira, R. M. N.Assuncao, M.Zeni, P. Polleto, *Ind. Crop. Prod.* **33**, 566 (2011).
7. Y. Li, Y. W. Mai, L.Ye, *Compos. Sci. Technol.* **60**, 2037 (2000).
8. G. Iñuguez-Covarrubias, S. F. Lang, R.M. Rowel, *Bioresour. Technol.* **77**, 25 (2001).
9. G. Iñuguez-Covarrubias, R.Diaz-Teres, R. Sanjuan-Dueñas, J.Anzaldo-Hernandez, *Bioresour.Technol.* **77**, 101 (2001).
10. L. Karla Tovar-Carrillo, S. Sugita-Sueyoshi, M.Tagaya, T.Kobayashi, *Industrial & Engineering Chemistry Research* (2013) (in press).
11. H. Kitano, K. Ichikawa, M. Ide, M. Fukuda, *Langmuir* **17**, 1889 (2001).
12. H. Kitano, M. Ide, T. Motonaga, *Langmuir* **22**, 2422 (2006).
13. J. A.Venegas-Sanchez, M. Tagaya, T. Kobayashi, *Ultrasonics Sonochemistry* **20**, 1081 (2013).
14. Y. Tamada, Q.Ikada, *J. Colloid. Interface. Sci.* **155**, 334 (1993)
15. A. K. Salem, S. J.Tendler, C. J. Roberts, *J. Biomed. Mater. Res.* **61**, 212(2002).
16. M. Kacurakova, R.H. Wilson, *Carbohydrate Polymers* **44**, 291(2001).

Mater. Res. Soc. Symp. Proc. Vol. 1613 © 2014 Materials Research Society
DOI: 10.1557/opl.2014.162

Molecular relaxation in Chitosan films in GHz frequency range

Siva Kumar-Krishnan, Evgen Prokhorov and Gabriel Luna-Barcenas

CINVESTAV-Queretaro, Libramiento Norponiente 2000, Queretaro, QRO 76230, MEXICO

ABSTRACT

The molecular relaxations behavior of chitosan (CS) films in the wide frequency range of $0.1\text{-}3\text{x}10^9$ Hz (by using three different impedance analyzers) have been investigated in the temperature range of -10^0C to 120^0C using Dielectric Spectroscopy (DS). Additionally to the low frequency molecular relaxations such as α and β relaxations, for the first time, high frequency (1-3 GHz) relaxation process has been observed in the chitosan films. This relaxation exhibits Arrhenius-type dependence in the temperature range of -10^0 C to 54^0C with negative activation energy -2.7 kJ/mol. At temperatures above 54^0C, the activation energy changes from -2.7 kJ/mol to +4.4 kJ/mol. Upon cooling, the activation energy becomes negative again with a value of -1.2 kJ/mol. The bound water between chitosan molecules strongly modifies molecular motion and the relaxation spectrum, giving rise to a new relaxation at the frequency at *ca.* 1 GHz. *In situ* FTIR analysis has shown that this relaxation related to the changes in vibration of the –OH, NH and –CO functional groups.

INTRODUCTION

The molecular motion and dielectric relaxation dynamics over high frequency range and the effect of moisture content in polysaccharide based biopolymer networks is an interesting and fundamentally important and not-well understood problem [1-3]. Dielectric spectroscopy (DS) is a powerful technique for the investigation of structure and dielectric relaxation phenomena and rotational dynamics of the molecules in polymers. It is noteworthy that, the physical and chemical properties of biopolymers can be significantly changed by the presence of small amounts of water [2]. The increasing numbers of studies of dielectric measurements in polysaccharides are controversially discussed about the shape of the dielectric spectra, number of relaxation process and their interpretation [4,5]. Among the main causes for the discussion are the very complex supramolecular structures of polysaccharides such as intra and intermolecular hydrogen bonds and the strong influence of water content [6]. The presence of water can significantly distort the results, giving rise to additional relaxation process in the material.

Dielectric relaxation studies on different polysaccharides have been reported in literature, mainly for cellulose [7] and starch [8]. Einfeldt et al [9,10] have reported the dielectric relaxations found in variety of polysaccharides that exhibit similar relaxation processes. Although a great deal of dielectric data exists for frequencies in the megahertz region and below, there exist a small number of articles about relaxation properties of polymers in the GHz range. Difficulties arise when measuring high resistivity polymer material with relatively low dielectric constant in the GHz range. Monitoring the temperature is also difficult due to convention such that thermocouples cannot be used in a microwave measurement. In this study, relaxation properties of chitosan films in the wide frequency range of $0.1-3 \times 10^9$ Hz (with help of three different impedance analyzers) was investigated in the temperature range -10^0C to 120^0C. Additionally, structural changes at the different temperatures have been investigated by *in-situ* Fourier transform infrared spectroscopy (FTIR).

EXPERIMENTAL DETAILS

Chitosan, 86% of degree of deacetylation (DD) and molecular weight of *ca.* 350 kDa was purchased from Sigma- Aldrich and was used as received. CS films were obtained by dissolving 1 wt % of CS in a 1 wt % aqueous acetic acid solution with subsequent stirring to promote dissolution. Films were prepared by the solvent cast method by pouring the solution into a plastic Petri dish and allowing the solvent to evaporate at 60°C. To obtain the neutralized films and to remove acetic acid, chitosan films were immersed into a 0.1M NaOH solution during 30 min and washed with distilled water until neutral pH; a subsequent drying step in furnace at 60°C for 14 hours was performed. A thin layer of gold was vacuum-deposited onto both film sides to serve as electrodes.

Dielectric spectroscopy measurements were carried out using three Impedance Analyzers: Solartron 1260 (in the frequency range $1-10^4$ Hz); Agilent 4249A (in the frequency range 10^2-10^7 Hz) and Agilent E4991A (in the frequency range $10^6-3 \times 10^9$ Hz). The high frequency measurements were carried out in the cell in the bench-top temperature chamber SU-261 with controlling temperature from -10^0C to 110^0C. This configuration has an advantage of providing high precision temperature control and impedance measurements over the entire frequency range. Supporting evidence was obtained from FTIR spectroscopy, (Perkin–Elmer) using an ATR accessory in the range 4000– 400 cm^{-1}; resolution was set to 4 cm^{-1}, with *in situ* temperature controller in the temperature range of 25°C to 110°C.

DISCUSSION

Typical dependence of dielectric loss ε" in the frequency range 1Hz to 3 GHz of as-prepared chitosan films (water content *ca.* 10 wt %) are shown on Figure 1 at the two selected temperatures (32°C and 53°C). These dependencies have been obtained on three Impedance Analyzers. These spectra have been calculated from dielectric spectroscopy measurements using DC correction as described elsewhere [1,10]. Figure 1 shows three relaxation processes: 1) low frequency α-relaxation; 2) β-relaxation (10^4-10^8 Hz) and 3) an additional new high frequency relaxation process in 1-3 GHz range. Here we principally concerned on the high frequency (GHz) relaxation behavior of CS films.

Figure 2, shows a typical dependence of the dielectric loss (ε″) with frequency of the as-prepared chitosan films at four different temperatures. The dielectric loss (ε″) shows a well-pronounced maximum at frequencies between 1-3 GHz. The fitting of the complex permittivity were carried out using the well-known Havriliak and Negami empirical model:

$$\varepsilon^* - \varepsilon_\infty = \frac{(\varepsilon_s - \varepsilon_\infty)}{\left[1 + (j\omega\tau)^\alpha\right]^\beta},$$

where ε_s-ε, and ε, are the dielectric relaxation strength and the dielectric constant at the high frequency limit, respectively; τ is relaxation time, the exponents α and β introduce a symmetric and asymmetric broadening of the relaxation peaks.

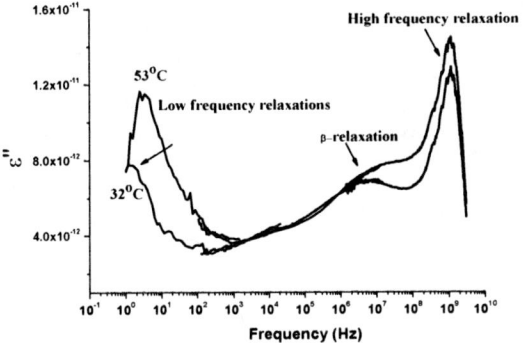

Figure 1. Dielectric loss factor (ε'') *versus* log frequency for chitosan films at the two indicated temperatures.

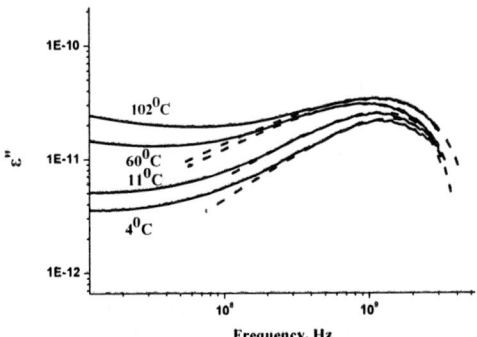

Figure 2. Dielectric loss factor (ε'') *versus* log frequency for chitosan film measured at the indicated temperature (continuous lines). Dashed-lines are results of fitting using Havriliak-Negami model.

As one can see from Figure 2, Havriliak-Negami empirical model fits well the high frequency relaxation process in CS films. Interestingly we observed a shift in the dielectric loss maximum to lower frequency until 50^0C, after that shifting to high frequency region.

Figure 3, shows dependencies of relaxation time *versus* reciprocal temperature for as-prepared CS films in the process of heating and cooling. Below 54^0C, the temperature dependence of relaxation time τ can be described by Arrhenius-type dependence: $\tau = \tau_0 \exp E_a / RT$, with negative activation energy $E_a = -2.7$ kJ/mol. Above 54^0C activation energy is positive and equal 4.5 kJ/mol. Negative activation energy have been observed in collagen tendon like structure [11] and can be related to the rotational motion of bound water molecules. The positive activation energy above 54^0C probably related to the broken of the hydrogen bonds between water molecules and NH, OH or CO groups of chitosan due to increasing thermal energy. The broken of strong hydrogen bonds increase in the mobility of the released molecules of bound water and this process is supported by positive values of the activation energy [12,13]. In the process of cooling, the activation energy again follow a negative (-1.2 kJ/mol) which is related to absorbance of moisture content in the process of cooling. Hoekstra and Doyle [14] pointed out that relaxation with low values of the negative activation energy observed in such frequency region (about of 1GHz) for Na-montmorillonite clay and γ-aluminum oxide films and related to surface absorbed water between two layers of the materials.

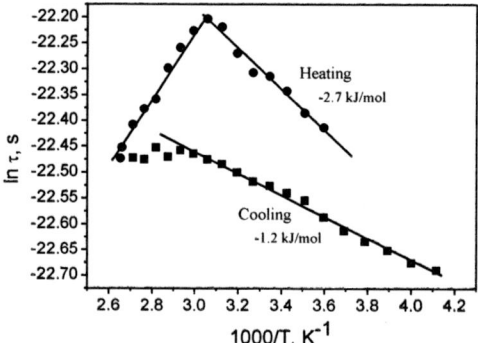

Figure 3. Temperature dependence of relaxation time *versus* 1/T for neutralized film. Lines represent Arrhenius-type dependence fitting.

Structural Analysis (FTIR)

Proposed mechanisms of interaction of the bound water molecules with CS obtained from dielectric spectroscopy measurements have been confirm by *in situ* FTIR measurements. Figure 4, shows a series of FTIR spectra of the film taken at different heating and cooling temperatures (analogously to dielectric spectroscopy measurements). As prepared CS film have absorption bands in the $3700-3000$ cm^{-1} frequency range arising from the stretching vibration of O-H, extension vibration of N-H, and the intermolecular hydrogen bonds of the polysaccharide and in the $1480-1750$ cm^{-1} range corresponded to the C=O stretching of secondary amides (amide I band) and amide II (N-H bending), respectively [15]. The bands at 1419 cm^{-1}, 1375 cm^{-1} and

1320 cm^{-1} corresponding to CH$_2$ bending, CH$_3$ deformation; CH bending and CH$_2$ wagging, respectively [16]. At the temperature higher 60^{0}C the intensity of N-H bending at 1590 cm^{-1} decreased, as well as the intensity of broad band at 3360 cm^{-1} corresponding to a stretching vibration of O-H groups decreased and shifted to 3450 cm^{-1}. Additionally, the intensive of the band at 1655 cm^{-1} decreases and shift to 1655 cm^{-1} assigned to C=O stretching [17], as well as C-H bending vibrations at 1300-1450 cm^{-1} decreases in intensity has been observed at the temperature higher 60^{0}C. This drop in intensity and shifts in absorption bands could be affecting the hydrogen bonded structure of the chitosan film due to dehydration of adsorbed water [6] and broken of the hydrogen bonds between water molecules and NH, OH and CO groups of chitosan.

In the cooling process (Fig. 4b), the intensity of O-H stretching vibration 3360 cm^{-1} and C=O stretching at 1655 cm-1 increased.

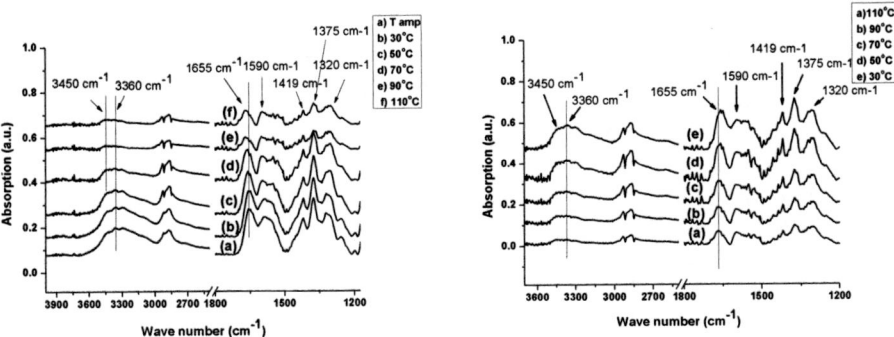

Figure 4. (a) FTIR spectra of chitosan films upon heating, and (b) upon cooling.

These changes are likely due to new types of hydrogen bonding in the swollen chitosan environment, which is indicative for chitosan behavior upon cooling compared with the dry polymer spectrum. It is noteworthy for films the large increasing of intensity of the band at 3000-3500 cm^{-1} during the cooling process; this observation can be traceable to the rapid moisture uptake of CS film [3]. So, in situ FTIR analyses have shown the changes in vibration of the –OH, -NH and –CO functional groups due to the change of the bound water molecules and confirm results of dielectric spectroscopy.

CONCLUSIONS

The high frequency dielectric relaxations behavior of chitosan films in the Gigahertz frequencies range was investigated. For the first time, a new relaxation process in the frequency range 1-3 GHz has been observed which attributed to the interaction of the interfacial bound water between two layers of the chitosan molecules. This relaxation exhibits an Arrhenius-like dependence in the temperature range of -10^0 C to 54^{0}C with negative activation energy -2.7 kJ/mol. At the temperatures above 54^{0}C, the activation energy changes from negative to positive values (+4.4 kJ/mol). *In situ* FTIR analysis has shown the changes in vibration of the –OH, -NH and –CO functional groups. The result of this study provides new insight into molecular motion and high frequency dielectric relaxation process in the chitosan.

ACKNOWLEDGMENTS

This work was partially supported by CONACYT of Mexico (Grant no. 181678). The authors are grateful to J.A. Muñoz-Salas for assistance in electrical and R. Araceli Mauricio Sánchez for assistance in FTIR measurements.

REFERENCES

1. J.B. Gonzalez-Campos, E. Prokhorov, G. Luna Barcenas, A. Mendoza-Galvan, I.C. Sanchez, S.M. Nuno-Donlucas, B. Garcia-Gaitan, Y. Kovalenko, *J. Polym Sci. Part B: Polym Phys.* **47**, 2259-2271 (2009).
2. M. Al Kobaisi, P. Murugaraj, D.E.Mainwaring, *J. Polym Sci Part B: Polym Phys.* **50**, 403-414 (2012).
3. M.T. Viciosa, M. Dionisio, J.F. Mano, *Biopolymer* **81**, 149-159 (2006).
4. T. Fukuda, A.Takada, T. Miyamoto, *In Cellulosic Polymers, Blends and Composites;* Gilbert, RD, Ed, Hanser: New York (1994).
5. H. Montes, K. Mazeau, J.Y. Cavaille, *Macromolecules* **30**, 6977-6984 (1997).
6. A.Nogales, T.A. Ezquerra, D.R. Rueda, M. Retuert, *J. Colloid Polym Sci.* **275**, 419-425 (1997).
7. D. Radloff, C. Boeffel, H.W. Spiess, *Macromolecules* **29**, 1528-1534 (1996).
8. M.F. Butler, R.E. Cameron, *Polymer* **41**, 2249-2263 (2000).
9. D. Maissner, J. Einfeldt, A.K. Wasniewski, *J. Non- Cryst. Solids* **320**, 40-55 (2003).
10. J. Einfeldt, D. Maissner, A. Kwasniewski, *Prog. Polym. Sci.* **26**, 1419-1472 (2001).
11. S.C. Harvey, P. Hoekstra, *J. Phys. Chem.* **76**, 2987-2993 (1972).
12. W.S. Rrey, J.T.E. Evans, L.H. Hitzrot, *J. Colloid and Interface Sci.* **26**, 306-316 (1968).
13. E. Marzec, L. Kubisz, F. Jaroszyk, *Int. J. Biological Macromolecules* **18**, 27-31 (1996).
14. P. Hoekstra, W.T. Doyle, *J. Colloid and Interface Sci.* **36**, 513-521(1970).
15. P. Murugaraj, D.E.Mainwaring, D.C.Tonkin, M. Al Kobaisi, *J. Appl. Polymer Sci.* **120**, 1307–1315 (2011).
16. E.S. Noriega, A. Subramanian, *Int. J. Carbohydrate Chem.* **2011**, 1155-1168 (2011).
17. J. Zawadzki, H. Kaczmarek, *Carbohydrate Polymers* **80**, 394-400 (2010).

Mater. Res. Soc. Symp. Proc. Vol. 1613 © 2014 Materials Research Society
DOI: 10.1557/opl.2014.163

Different Materials of Substrates in the Production of Chili Apple Seedlings (*Capsicumpubescens* R. and P.) Grajales ST

Martha L. Domínguez P.[1], Oscar G. Villegas T.[2], Héctor Sotelo N.[2], Carlos M. Acosta D.[2], Mario Pérez G.[3], Diana Rodríguez B.[2]
[1] Faculty of Chemical Sciences and Engineering. University Autonomous of the State of Morelos, Cuernavaca, Morelos 62209, Mexico.
[2] Faculty of Agricultural Sciences. University of the State of Morelos, Cuernavaca, Morelos 62209, Mexico.
[3] Institute of Horticulture. University Autonomous of Chapingo. Chapingo, 56230, Mexico.

ABSTRACT

This study aimed to evaluate the effect of 12 substrates in the seedling growth of chili apple (*Capsicum pubescens* R. & P.) hybrid Grajales ST. The substrates were prepared with different proportions of five materials: perlite, coconut fiber, loam, Sunshine3 ® and wood dust. The seeds were sown in trays of 50 cavities. For the substrates were determined physical and chemical properties. Significant differences in growth parameters of seedlings are the effect of the substrate. Based on the remarkable accumulation of dry matter of each of the organs of the seedling, and their relative distribution, one can surmise a seedling quality with 47.70 % dry matter in leaves, 35.34 % in stem and 16.95 % in root. The substrate with better features for the production of chile apple seedlings was prepared with 25 % loam and 75 % perlite (v/v), which showed retention of 85.5 % moisture, electrical conductivity of 0.03 $dS \cdot m^{-1}$ and total porosity of 62.5 %.

INTRODUCTION

There are vegetables whose demand is low due to low dietary inclusion of consumers, but there are others that are consumed mainly for cultural reasons, as in the case of chili apple (*Capsicum pubescens* R. and P.) who is originally from the highlands of Peru and Bolivia.

In some European country clubs and the United States, the demand is to cover the consumption of part, because its population consists mainly of immigrants. Besides adding flavor to food, chili has nutritional qualities for its vitamin A and C [1]. The chili is a culinary ingredient that identifies Mexicans and represents an important value from the point of view of agriculture, economic and social. Their use is indispensable in areas of the mountain north of Puebla, Mexico and Michoacan state, where it forms part of the eating habits.

In the past five years, the field observations demonstrate the existence of at least 1500 ha, distributed in the mountain north of Puebla, Tacambaro and Zitacuaro in Michoacan, Veracruz Huatusco and Zongolica, Coatepec San Miguel Flour and Tlaixpan in the State of Mexico and San Cristobal of Houses in Chiapas Motozintla, among other places [2]. One of the demands of intensive systems is the use of seedling produced under controlled conditions, because this way the same quality is superior to those produced in the field. Studies show a direct relationship between the quality of the seedling with the fruits and performance [3-6]. Among the factors affecting the quality of the seedlings are container size, age transplant, nutrition, and the type of substrate [3, 7, 8]. The solid substrate is any different from the ground, can be natural or synthetic, mineral or organic, and placed in container, pure or mixed, can anchor the plants through the root system or the substrate may act not in the process of plant nutrition. The latter classifies chemically inert substrates (perlite, rockwool, volcanic rock) and assets (peat, pine bark, etc.) [2]. Growing plants in substrate substantially are different compared with those produced in soil. When growing in containers, its characteristics are crucial in the proper growth of the plant, due to the interaction between its characteristics (height, diameter, etc.) and plant-substrate complex [10]. The characteristics of the substrates have been modified to obtain better results in the production of container plants [11] so it is important to know their physical-chemical and biological characteristics [12]. The physical properties of a substrate, the total porosity, the water retention capacity, the actual and apparent density are those that determine the potential for a material to be used as substrate, while the chemical properties of a medium are important because of their interaction with fertilizers and their effect on plant growth, the pH and electrical conductivity in determining the selection of the substrates [13]. A good substrate should have high porosity (85%), of which, a good ratio should be macro pores, and at least 50% of water retention [12].

In tomatillo, best led emergency peat and coconut fiber it was for seedling growth compared with the results obtained with pig vermicompost [14]. Magdaleno [15] evaluated the effect of coconut fiber, peat and vermicompost plastic and two colors (black and silver) on the rate of emergence and growth of seedlings of tomato peel. Indoor use was not relevant in the production of tomato seedlings under ambient conditions, and the best substrates for the emergence and seedling development were peat and coir; vermicompost seedlings produced lower quality. The aim was to evaluate the effect of 12 substrates in the development of the seedlings of chili apple (Capsicum pubescens R. and P.) hybrid Grajales ST.

EXPERIMENTAL DETAILS

The work was conducted in a tunnel covered with transparent polyethylene 50% shade cloth, located in the experimental field (18 ° 58 '53" LN, 99 ° 13' 58.4" LO, 1804 m) of the Faculty of Science Agricultural Autonomous University of Morelos State, Campus Chamilpa.

The climate of the region is A (C) w" 1 (w) ig, semiannual mean temperature between 18 and 22 °C, the percentage of winter precipitation relative to the annual total is less than 5 mm, in the presence of heat wave , annual variability in monthly mean temperatures below 5 ° C and the temperature up type Ganges [16].

Chilli seeds were used apple Chapingo Amarillo hybrid is a cross between the varieties Zongolica x Puebla. The plant is perennial with short (11 cm) and middle leaves, produces a large number of fruits, yellow, average weight of 52 g with two to three cores. The first time after transplantation ripe fruit is five months [2].

The seed was soaked in water for 24 h to soften the seed coat and promote uniform germination. It planted a seed per well in black polyethylene trays 50 cavities containing any of the substrates tested (Table 1). The seeds were watered with a solution of Captan® (1 g · L^{-1}) and then the trays were covered with a plastic sheet for a period of 7 d, time when kept in a seeder, at a temperature between 22 to 25 ° C.

Then the trays were taken to the greenhouse where the seedlings continued its development until the time of the evaluation, which was when they presented the characteristics for transplantation. During this time, there were two watering a day, depending on ambient conditions.

From the trays were placed in the greenhouse and to seedling emergence, were watered with tap water. When presented 50% of the emergency and even visual appearance of the true leaves, the nutrient solution was used universal Steiner [17] to 30% of the original concentration.

The experiment consisted of 12 treatments for each given one of the substrates (Table 1). It was developed with a completely randomized design with three replications. The experimental unit consisted of 25 plants produced in black polyethylene tray cavities fifty. Each repetition seedlings took five full competences for evaluating the response variables. Each of the substrates was determined bulk density and true density, the percentage of moisture retention, total porosity, temperature, pH.

Table 1. Substrates used to evaluate the development of hybrid apple seedling Chapingo Amarillo chili.

Substrate	Composition
(Th)	100 % organicsoil
(S3)	100 % Sunshine3 ^{R}
(Fc)	100 % coconutfiber
(S3/Fc)	50 % Sunshine3 ^{R}-50 % coconut fiber (v/v)
(Th/As)	50 % organic soil -50 % of sawdust (v/v)
(Fc/As)	50 % coconutfiber -50 % of sawdust
(S3/As)	50 % Sunshine3 ^{R}-50 % of sawdust
(S3/Ag)	50 % Sunshine3 ^{R}-50 % agrolita (v/v)
(Th/Ag)	25 % organic soil -75 % agrolita (v/v)
(S3/As)	75 % Sunshine3 ^{R}-25 % of sawdust (v/v)

(Fc/As)   80 % coconut fiber -20 % of sawdust (v/v)

(S3/Ag)   75 % Sunshine3®-25 % agrolita (v/v)

* The Sunshine3 ® composition is: 70% to 80% Enfagnacea Canadian peat, vermiculite, limestone (for pH adjustment), agricultural gypsum and a wetting agent.

To determine the physical and chemical characteristics of substrates used by the methodologies it described respective [18].

Percent porosity. To one liter of dry substrate at room temperature water was added until saturation (mirror formation water). The relationship between the water used for the saturation of the substrate and the initial volume of water (1 L) corresponded to a percentage pore   space.

Real density. It weighed a liter of dry substrate and subtracted the total pore space.

Bulk density. Gravimetrically determined considering the weight of 1 liter of dry substrate.

Temperature. This was determined in the substrate container capacity introducing the thermometer bulb (Taylor Bi, Therm) to the center of the substrate.

pH was measured with portable potentiometer (Rapistet) in the substrate container capacity by placing the sensor at the center of it.

Electrical conductivity was determined with the conductivity (SR 993310) in the leachate obtained after saturation of the substrate.

Moisture retention capacity. Water was applied to one liter of dry substrate at room temperature until saturation, then allowed to leach excess and when it stopped was determined drain moisture retention capacity subtracting the initial volume (1 L) volume leachate.

DISCUSSION

Seedling characteristics varied depending on the type of substrate or mixture. The best substrates for chlorophyll content were S3/Ag seedlings (50,50) and Th (100) that exceeded 53.46 and 49.26% respectively as obtained in seedlings grown in the Fc substrate (100) which responded least in this variable (Table 2). It was observed that when it have been used sawdust in combination with any principal components, it generated seedlings with low content of chlorophyll (31.78-36.06 units SPAD), which could have been due to release phenols from sawdust. The overall average was 36.03 SPAD units.The average was also surpassed byS3/Ag grown seedlings (50,50) and Th (100), 15.68 and 12.51%, respectively. This result may be explained because in the treatment S3/Ag (50,50) and Th (100), the electrical conductivity exceeds 50 and 100% Fc treatment (100).

The highest value of dry matter was obtained root seedlings grown in Th / Ag (25,75), probably due to better absorption of water and nutrients, the lowest value was in those grown in Fc (100) . In total dry weight, the highest value was in seedlings grown in Th / Ag (25.75), and lowest in the developed Fc (100) (Table 3). Dry matter weight of leaves, stem, root and seedling planted of Th / Ag (25, 75) was increased. This behavior may indicate a contribution of nutrients to achieve optimal growth of seedlings with reduced root systems. In the biomass production is important to consider that the different plant organs have a fundamental role in the production crop [19].

Table 2. Physical and chemical properties of the substrates used for the production of hybrid apple seedlings chili apple Grajales ST.

Substrates	DA	DR	T	pH	CE	PT	RH
Th100	362	1448	20	7.0	0.16	75.0	57.0
S3100	133	887	20	7.0	0.30	85.0	70.0
Fc100	86	860	19	7.0	0.12	90.0	63.0
S3/Fc(50,50)	108	600	21	6.9	0.12	82.0	66.5
Th/As(50,50)	252	008	22	7.0	0.12	75.0	61.0
Fc/As(50,50)	103	412	21	6.9	0.22	75.0	53.5
S3/As(50,50)	127	508	21	6.9	0.21	75.0	61.5
S3/Ag(50,50)	156	918	21	7.1	0.08	83.0	60.5
Th/Ag(25,75)	344	917	20	7.0	0.03	62.5	85.5
S3/As(75,25)	123	230	21	7.0	0.49	90.0	82.5
Fc/As(80.20)	920	613	20	6.9	0.13	85.0	64.5
S3/Ag(75,25)	143	867	21	7.0	0.20	83.5	63.5

DA, bulk density (g • L-1), RD, true density (g • L-1), T is temperature (° C) EC, electrical conductivity (dS m-1), PT, total porosity (%); RH, moisture retention (%) Th, leaf mulch, S3, Sunshine3 ®, Fc, coconut fiber, As, sawdust, Ag, agrolita.

CONCLUSIONS

The substrate with better features for the production of chili apple seedlings in trays of 50 cavities was Th / Ag (25,75) which showed 85.5% retention of moisture, electrical conductivity of 0.03 dS m^{-1} and 62.5% porosity overall, although other substrates, seedlings also showed favorable characteristics: Th (100), S3/As (50,50), S3/Ag (50,50) and S3/As (75.25), with margins of 57.0 -82.5% moisture retention, electrical conductivity from 0.08 to 0.49 dS m^{-1} and 75.0-90.0% total porosity.

REFERENCES

1. J. V. Maroto, *Horticultura herbácea especial* (Mundi-Prensa, 2002) pp. 120-125.
2. M. Pérez y B. Castro, *El chile manzano*, (Universidad Autónoma Chapingo, 2009) pp. 10-15.
3. L. A. Weston and B. H. Zandstra, *Hort Science* 24, 88-90 (1989).
4. V. Markovic, M. Djurovka, and Ž. Ilin, *Acta Horticulturae* 462, 163-167 (1997).
5. T. Taga, *Técnicas de trasplante en el cultivo de tomate* (Centro Nacional de Tecnología Agropecuaria y Forestal, 2003) pp 1-5.
6. S.Saldaña *Producción hidropónica y transformación agroindustrial del tomate saladette* (*Lycopersiconesculentum*). Curso precongreso. XVI Congreso Nacional de Ingeniería Bioquímica, V International Biochemical Engineering, VI Jornadas Científicas de Biomedicina y Biología Molecular. Tuxtla Gutiérrez, Chiapas (2008).
7. S. Nicola and L. Basoccu, *Acta Horticulturae* 361, 519-526 (1994).
8. P. Cornillon, *Acta Horticulturae* 487, 133-137 (1999).
9. J.N. Pastor, *Terra* 17, 231-235 (1999).
10. M. Abad, P. Noguera. *Los sustratos en los cultivos sin suelo*, editado por M.Urrestarazu (Mundi-Prensa, 2000) pp 137-184.
11. M. Andrade-Rodríguez, J. J. Ayala-Hernández, J. Arce, C. M. Acosta-Durán, I. Alia-Tejacal, V. López-Martínez, O. G. Villegas-Torres, *Revista Investigación Agropecuaria* 4, 9-16 (2007).
12. F. R. Díaz-Serrano, *Selección de sustratos para la producción de hortalizas en invernadero*. Memorias del IV Simposio Nacional de Horticultura. Invernaderos: Diseño, Manejo y Producción. Torreón, Coah, México (2004).
13. G. Quesada, C. Méndez. *Agronomía mesoamericana* 16, 171-183 (2005).
14. A. Peña, J. Magdaleno, R. Mora, P. Becerra, *Evaluación de sustratos y plásticos para la producción de plántula de tomate de cascara* (*PhysalisixocarpaBrot.*). VIII Congreso Nacional Agronómico. Universidad Autónoma Chapingo (2005).
15. J. J. Magdaleno, A. Peña, R. Castro, A. M. Castillo, A. Galvis, F. Ramírez, P. A. Becerra, *Revista Chapingo Serie horticultura* 12, 153-158 (2006).
16. E. García, *Modificaciones al sistema de clasificación climática de Köppen (para adaptarlo a las condiciones de la República Mexicana)* (Larios, 1981) México. pp. 251
17. A. A. Steiner, *The universal solution*, Proceedings of 6th International Congress on Soilles. Culture. Lunteren, The Netherlands pp. 633-649 (1984).
18. J. Ansorena, *Sustratos. Propiedades y caracterización* (Mundi-Prensa, 1994) pp. 172
19. M. Peil, J. L. Gálvez, *Agrociencia* 11, 5-11 (2005).

Mater. Res. Soc. Symp. Proc. Vol. 1613 © 2014 Materials Research Society
DOI: 10.1557/opl.2014.164

Application of Glass Particles Doped by Zn^{+2} as Antimicrobial and Atoxic Compound in LLDPE and HDPE

Marcel Ferrari dos Santos[1]; Camila Machado de Oliveira[1]; Camila Tachinski[1]; Jair Fiori Júnior[2]; Raquel Piletti[2], Clever Pirola Ávila[3], Thiago Tófoli[3], Jaqueline Madeira Gonçalves[1] and Marcio Antônio Fiori[1]

[1] Universidade do Extremo Sul Catarinense – UNESC, Brazil.

[2] Universidade Federal de Santa Catarina - UFSC, Brazil.

[3] Marfrig Group, Brazil.

ABSTRACT

This study demonstrates the potential application of glass particles doped with Zn^{+2} (GZn) as antimicrobial additives and atoxic of the HDPE and LLDPE polymers. Toxicity tests indicated the absence of toxicity in human cells. Microbiological tests proved the antimicrobial effect of GZn pure compound and of the additives polymeric compounds (HDPE/GZn and LLDPE/GZn). Have also indicated that with percentages of GZn higher than 2.00 wt% and a time of 4 hours the bactericidal performance is excellent and equal for both polymeric compounds.

INTRODUCTION

The global consume of polymers has increased in the last decades, from 5 million of tonnes in 1950 to 100 million of tonnes in the current decade. Approximately 42.0% of this volume is destined for the manufacture of packaging of Low Density Polyethylene - LLDPE and High Density Polyethylene –HDPE [1].

The LLDPE and HDPE are polymers applied in the manufacture of active packaging for foods. The active packaging have antimicrobial properties and/or antioxidants and are used with the main objective to extend the shelf life of foods and increase the safety for the consumer [2-3].

Different organic and inorganic antimicrobial agents have been employed in the manufacture of the active packaging [4-5] and in studies with polymeric materials [6-7]. Inorganic compounds of silver, gold, titanium, copper and zinc are studied and applied to polymeric materials due to its antimicrobial activity [8]. However, some metals like silver and gold have restrictions of use due to its high cost and toxicity [9]. But zinc compounds have attracted interest as antimicrobial additive to be generally atoxic, low cost and high efficiency antimicrobial [10-11].

This study aims to evaluate the application of vitreous atoxic compounds doped with zinc ionic species (GZn) as antimicrobial additive of polymer. The work explores the application of the GZn compound as an additive of HDPE and LLDPE. The results show the characterizations of toxicity, microbiological and capacity release of ionic zinc species of the GZn particles and by the polymer matrices.

EXPERIMENTAL DETAILS

The polymers used in this work were the Linear Low Density Polyethylene - LLDPE (Dow Chemical Company) and High Density Polyethylene - HDPE (Braskem). The vitreous compound doped of zinc ion, denominated by GZn, was produced by Kher and Chemical Research supported by the Laboratory for Advanced Materials and Processing - LMPP, following the procedures of Fiori et al. [12].

The production of compounds LLDPE/GZn and HDPE/GZn was realized with the incorporation of the composed GZn in the polymer matrices of LLDPE and HDPE with different percentages, ranging from 0.0 to 2.4 wt%. The homogenization was performed in a single screw extruder, model Oryzon OZ-E-EX-L22, L/D 17, with controlled temperatures and the screw rotation speed of 90.0 rpm.

The vitreous composite doped with zinc ions (GZn) and pellets of LLDPE/GZn and HDPE/GZn were submitted to toxicity tests of comet type and microbiological tests of type Agar Diffusion and Death Curve, test with *Escherichia coli* (ATCC 8739) and *Staphylococcus aureus* (ATCC 25923). The tests were conducted following the procedures established by international standards [13-14].

DISCUSSION

Toxicity Analysis

Figures 1a and 1b show the results of toxicity tests of the type comet for compound GZn. The answers are the index of fragmentation and the frequency of DNA damage in human blood cells, respectively.

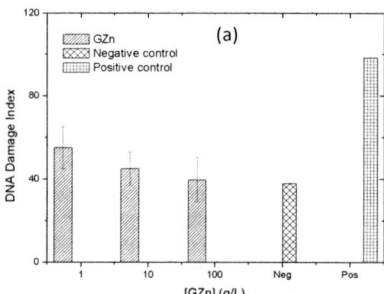 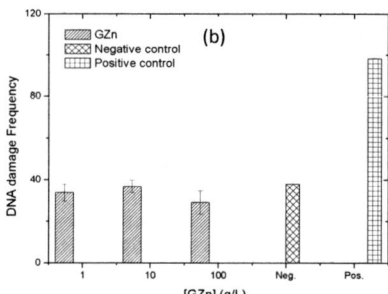

Figure 1. Results of toxicity tests of comet type for compound GZn. (a) Fragmentation index and (b) damage frequency to the DNA humans.

The fragmentation index values didn't show statistically significant differences with increasing concentration of GZn. The average value for the fragmentation index is 46.6 ± 7.8, higher than the fragmentation index of the negative control, which is approximately 38.0. The results of damage frequency, figure 1b, show that the damage frequency doesn't depend on the concentration of GZn and has an avarage value of 33.2 ± 3.9, less than the negative control which also is 38.0.

In all tests the index of fragmentation and damage frequency showed lower values than the positive control. These results confirm the characteristic atoxic of compound GZn at the concentrations tested, even at concentrations higher than 100 g/L. The GZn particles are composed of vitreous material that is atoxic. Even with the release of ionic zinc the toxicity is low, because studies show that compounds consisting of zinc have low toxicity [15-16].

Antimicrobial Analysis

Figure 2 shows the results microbiological of agar diffusion with bacterias of type *Staphylococcus aureus (SA) and Escherichia coli (EC)* for the GZn. The results indicate the presence of antimicrobial effect on both types of bacteria. The presence of zones of inhibition indicates the presence of oligodynamic effect of the ionic zinc species with inhibition of bacterial growth.The antimicrobial effect may be attributed to the ionic zinc species present on the surface of the GZn particles or of the ioniczinc species are released to the external medium of the vitreous particles. In both situations can occur inhibiting the growth of the bacterial.

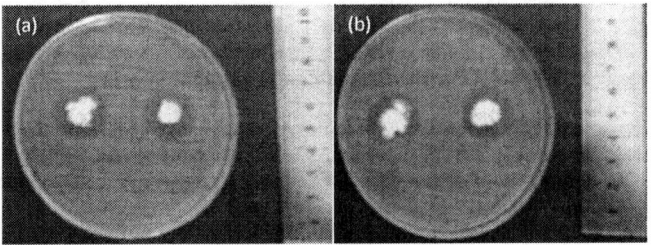

Figure 2. Microbiological results of agar diffusion for (a) GZn with *Staphylococcus aureus* (SA) and (b) GZn with *Escherichia coli* (EC).

From the results of agar diffusion is possible to say the GZn has a bacteriostatic effect by having large inhibition zones with high antimicrobial capacity. Thus, the compound GZn presents a high antimicrobial effect and default of toxicity. Its effect on gram negative and gram positive bacterias enables its application as an additive in a large spectrum of type of bacteria. Its incorporation will aggregate the antimicrobial property to Linear Low Density Polyethylene (LLDPE) and High Density Polyethylene (HDPE).

Figure 3 shows the results microbiological realized with the HDPE/GZn and LLDPE/GZn samples containing different percentages of GZn. The tests evaluate the number of bacteria colonies surviving in contact with the polymers after 4 hours. The results don't show statistically significant differences for antimicrobial performance after 4 hours. For both polymeric compounds the reduction of bacterial colonies after 4 hours of contact was approximately 70.0% with the bacteria E.C. and approximately 95.0% with the bacteria SA, Figure 3b.

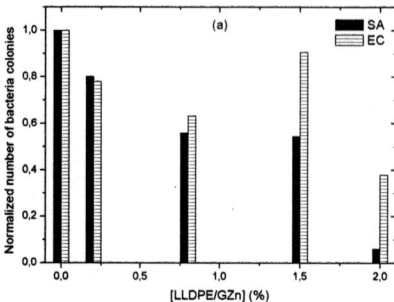 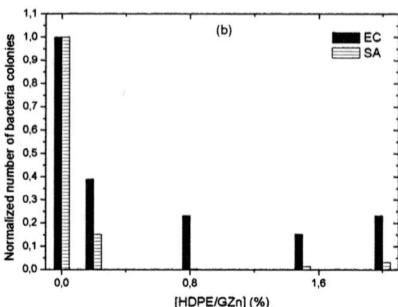

Figure 3. Amount of bacterial colonies surviving after 4 hours of contact with antimicrobial polymer samples as function of the GZn percentage in the polymer matrix. (a) LLDPE/GZn and (b) HDPE/GZn.

The results suggest that the amount of Zn^{+2} ionic species responsible for the antimicrobial effect by the polymer composite HDPE/GZn is practically the same regardless of the percentage of GZn in the HDPE matrix. With percentages below 0.5 wt% the compound HDPE/GZn now reduces 60% the number of bacterial EC colonies and 85% of the number of SA colonies. Thus, it is likely that the percentage of GZn in HDPE matrix doesn't affect significantly the antimicrobial effect.

In turn, with percentages of GZn below 0.5 wt% in the polymer matrix of PEBDL the composite LLDPE / GZn eliminates only 20% of the colonies of EC and SA bacteria. Still, by increasing the percentage the colonies number is reduced to equivalent amounts to the compound HDPE / GZn.

These characteristics suggest that the percentage of GZn not affect the performance of HDPE / GZn but positively affects the antimicrobial performance of LLDPE / GZn. Even after 4 hours both antimicrobial compounds achieve equivalent antimicrobial performance.

CONCLUSION

This work proves that the compound GZn has characteristics not toxic to human cells and can be used with safety as antimicrobial additive to polymers. Microbiological tests confirmed that the compound GZn has antimicrobial effect in bacteria of the type Gram positive and Gram negative.

When incorporate the GZn in the HDPE and LLDPE aggregates the antimicrobial property in the polymeric compounds. After 4 hours and with percentages of GZn from 2.0 wt%

both compounds HDPE/GZn and LLDPE/GZn exhibit excellent antimicrobial performance practically equal.

REFERENCES

1. C. Silvestre, D. Duraccio and S. Cimmino, *Progress Polymer Science* **36**, 1766 (2011).
2. A. M. Bonilla and M. F. Garcia, *Progress in Polymer Science* **37**, 281 (2012).
3. P. Singh,A. A. Awani and S.Saengerlaub, *Packaging of Food Products* **41 (4)**, 249 (2011).
4. J. E. Bruna, A.Penaloza, A.Guarda, F. Rodríguez and M. J. Galotto, *Applied Clay Science* **58** 79 (2012).
5. A. Llorens, E. Lloret,P. A. Picouet,R. Trborjevich and A. Fernandez, *Trends in Food Science & Technology* **24 (1)**, 19 (2012).
6. I. P. S. Thomé,V. S. Dagostin,R. Piletti,C. T. Pich,H. G. Riella,E. Angioletto and M. A. Fiori, *Material Science Engineering C* **32**, 263 (2011).
7. L. M. L. Toms,M. Allmir,J. F. Mueller,M. Adolfsson-Erici,M. Mclachlan, J.Murby and F. A. Harden, *Chemosphere* **85**, 1682 (2011).
8. M. M. S. Paula, C. V. Franco, M. C.Baldin, L. Rodrigues, T.Barichello, G. D.Savi, L. F.Bellato, M. A.Fiori and L. Silva, *Materials Science and Engineering C* **29**, 647 (2009).
9. R. Wahab,A. Mishra,S. Yun,I. H. Hwang,J. Mussarat,A. A. Al-khedhairy, Y. Kim and H. Shin, *Biomass and Bioenergy* **39**, 227 (2012).
10. Y. Xie,Y. He, P. L. Irwin, T. Jin and X. Shi, *Applied and Environmental Microbiology*, **77 (7)** 2325 (2011).
11. M. J. Salgueiro, M.Zubillaga, A.Lysionek, M. I.Sarabia, R. Caro, T.Paoli, A.Hager, R. Weilland and J.Boccio, *Nutrition Research*, **20 (5)**, 737 (2000).
12. ISO 20645:2004 (International Organization for Standardzation). Textile Fabrics: Determination of Antibacterial Activity – Agar Diffusion Plate Test.
13. JIS Z 2801:2000 (Japanese Industrial Standard). Antimicrobial Products – Test for Antimicrobial Activity and efficacy.
14. Y. Kao,Y. Chen,T. Cheng,Y. Chiung and P. Liu, *Toxicological Sciences* **125 (2)**, 462 (2011).
15. T. K. Sontakke,R. N. Jagtap,A. Singh and D. C. Kothari, *Progress in Organic Coatings* **74**, 582 (2012).
16. S. P. Denyer, and J. Y. Maillard, *Journal of Applied Microbiology* **92**, 35 (2002).

Polymer Characterization

Mater. Res. Soc. Symp. Proc. Vol. 1613 © 2014 Materials Research Society
DOI: 10.1557/opl.2014.165

On-Line Calorimetry in the Ethylene Coordination Polymerization

José R. Infante-Martínez, Enrique Saldívar-Guerra, Odilia Pérez-Camacho, Víctor Comparán-Padilla, Maricela García-Zamora

Centro de Investigación en Química Aplicada, Blvd. Enrique Reyna #140, Saltillo, Coahuila, 25294 México

ABSTRACT

The kinetic performance of metallocene type catalysts as well as their instantaneous activity is determined on line by two independent methods in the semi-batch polymerization of ethylene via metallocenes. On the basis of first-principles, both methods are described and guidelines for their implementation at a laboratory scale reactor are offered. Polymerization tests were conducted with two heterogenized metallocene catalysts showing that the direct method (based on ethylene flow measurement) and also the calorimetric method (based on energy balances) reported equivalent high quality information. The calorimetric method here developed can be readily used by the chemical practitioner as the notions and tools required for its implantation are easily grasped. It is noted that the calorimetric method has the advantage of requiring a low cost instrumentation (only thermocouples) whereas the direct method needs a relatively more sophisticated equipment (mass flow meter).

INTRODUCTION

The activity of a Ziegler-Natta type catalyst, defined as the polyethylene weight produced per mol of catalyst and per unit time, is an important technological variable that expresses the performance of a catalyst in the operation of an industrial plant of polyethylene. Frequently, mainly in industrial practice, mean activity values are used when comparing different catalysts, however this information is not enough, because one same mean activity value can result from widely different kinetic behavior. For instance, a multi-site catalyst could have the same mean activity as a single site catalyst [1]. Instantaneous activity values are a better choice to characterize a given catalyst. These are determined experimentally in low scale polymerization tests that report the instantaneous polymerization rate (kinetic curve). The evolution of the polymerization rate, r_p, can be considered the fingerprint of the catalytic system. Additionally to its value as a catalyst characterization means, a kinetic curve can guide the modelling studies, required for a deep understanding of the catalytic medium [2,3].

Usually, the metallocene (or Ziegler-Natta) catalyst behavior study is performed in a semi-batch stirred tank reactor provided with a heat transfer jacket, operating at the 0.2 L to 2 L scale of operation. The pressure is maintained constant by means of a pressure regulator. In this way, the addition of ethylene is continuous and governed by the reaction requirements. In the measurement of this flow directly provides the instantaneous polymerization rate. The reactor temperature (or jacket temperature) is automatically controlled. In the simplest case, the jacket temperature is maintained constant by means of a controlled temperature circulator connected to the jacket.

An alternative way to the direct ethylene flow measurement for the estimation of the catalytic coordination polymerization rate is through the estimation of the instantaneous heat of reaction. This method (calorimetric) has been employed successfully in a variety of polymerization reactions [4-9]. In a recent report [9], the calorimetric method has been used for the monitoring of ethylene polymerization via metallocenes in a laboratory reactor. The authors refer to successful results when they develop a calorimetric observer and compare it with an estimation of ethylene consumption based on the pressure of an ethylene reservoir.

As the instrumentation required to implement a calorimetric monitoring technique on or off-line (that is: at the moment of the test or a posteriori) is standard and economically accessible to most laboratories (only temperature sensors in the reactor and the jacket are necessary), it is convenient to consider the applicability of the calorimetric method to the reaction rate monitoring of the ethylene coordination polymerization and compare it with the most standard monitoring method (direct measurement of ethylene flow). The implementation of the calorimetric method is developed using notions and tools readily accessible to the chemical practitioner.

The study is conducted in a bench scale polymerization reactor equipped for the ethylene flow measurement as well as for the reaction heat evolution estimation. Two metallocene type catalysts are used that illustrate the applicability of the kinetic monitoring techniques. It is also demonstrated how these kinetic studies provide essential information for a deeper understanding of the chemical and dynamic nature of the catalytic system and to compare different catalytic systems based on their kinetic performance.

EXPERIMENTAL DETAILS

The experimental arrangement employed is shown in figure 1.

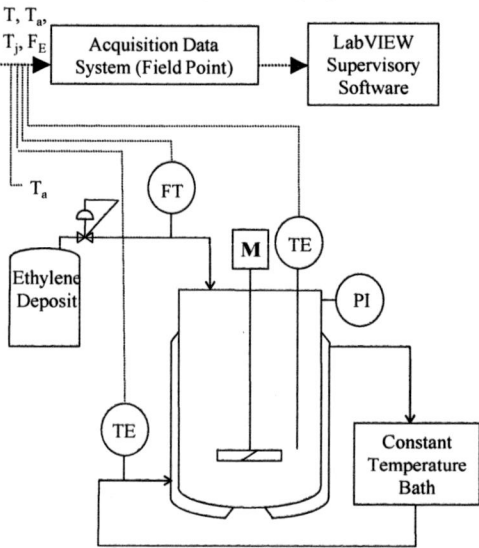

Figure 1. Laboratory reactor arrangement for semi-batch ethylene polymerization using metallocene catalysts.

104

It consists of a metallic-top, 600 mL glass PARR reactor provided with a glass jacket for heat exchange. The agitation is provided with a magnetically sealed mechanic drive. The reactor operation can operate at vacuum or light pressures (< 150 psig) at 150°C or lower temperatures.

A constant temperature circulating bath is connected to the reactor jacket. The temperature measurements are taken with type T thermocouples.

Ethylene Flow Measurement

The ethylene flow entering the reactor is measured via an Aalborg gas flow-meter of 0 - 1,000 mL/min flow range. The measuring principle of this instrument is based on the heat capacity determination of a small fraction of the total stream flowing through the apparatus. This heat capacity is directly related to the mass flow of gases flowing through the capillary and, by means of standardized constants, the ethylene mass flow can be obtained.

Data Acquisition System

It is a National Instruments Field Point 2000 bank, consisting of an Ethernet module containing a microprocessor with associated peripherals; a thermocouple module; and an analog input signal module (configurable as current or voltage inputs). The signals proceeding from the process (temperatures and ethylene flow) are transmitted to the data acquisition system where they are digitalized and directed to the supervisory control system (PC computer running LabVIEW software) as depicted in figure 1.

Supervisory Control System

The supervisory control system implemented is based on the National Instruments LabVIEW 9 Software. It runs on a conventional personal computer with MS Windows environment. Conceptually, it constitutes the interface between the operator and the process. The operator actions and the process responses are performed and visualized through the supervisory control system. Its main functions are: i) Communication with the data acquisition system to update in real time all the variables of the process; ii) Display of the main variables of interest by means of windows and graphical objects; iii) Register of the process variables in the computer hard disk.

Ethylene polymerization via metallocenes

Homogeneous and heterogeneous ethylene polymerizations were carried out in the 600 ml semi-batch reactor, equipped as described above in figure 1. The bath temperature was maintained at 70 ± 1 °C. Other reaction conditions used were: 42 psig of ethylene pressure and a stirring rate of 500 rpm. 200 ml of isooctane as solvent were used.

DISCUSSION
Determination of r_p through calorimetry
The basis for the method to follow the reaction rate through the reaction heat evolution is an energy balance applied to the reactor volume. The reactor is of the stirred type, so it has homogeneous conditions in all the volume of the reaction mixture.

The energy balance, with the standard terms of inlet (In), outlet (Out), production (Prod), and accumulation (Acc) is as follows [10]:

$$\text{In - Out + Prod = Acc} \tag{1}$$

For the laboratory setup (depicted in the experimental section):

$$-F_e c_{pe}(T-T_e) - UA(T - T_j) + r_p V(-\Delta H)M_e = \frac{d\ m_R C_{pR} T}{dt} \tag{2}$$

where F_e is the ethylene flow, g/s; c_{pe} and C_{pR} are the specific heats of ethylene and the reaction mixtures respectively, cal/g-°C; ΔH is the heat of reaction, cal/mol; T, T_e, and T_j are the reaction, ethylene inlet, and jacket temperature respectively, °C; UA is the product of heat transfer coefficient and the heat transfer area, cal/s-°C; r_p is the polymerization rate, mol/s-L; M_e is the molar mass of ethylene, g/mol; V is the volume reactor, L; t is time, s.

Equation (2) can be simplified as the inlet and accumulation terms are one or two orders of magnitude smaller than the outlet and production terms. The assumption that the accumulation term is zero (quasi-steady state assumption) introduces a slight error when the temperature experiences abrupt changes (dTd/dt >> 0) and will be minimum when these changes are smooth.

An advantage of this approximation is that less data treatment is required as we are riding off a signal that introduces inaccuracies in the calculations (noise). Otherwise data filtering or additional adjustments would be required.

The energy balance reduces to:

$$UA(T - T_j) = r_p V(-\Delta H)M_e \tag{3}$$

The polymer weight produced can be precisely known at the end of the test, so we can define a calibration constant that includes all the inaccuracies of the method. Using Equation (3):

$$r_p = \frac{UA}{V(-\Delta H)M_e}(T - T_j) \tag{4}$$

Integration of Equation (4) along the reaction time yields the mass of polymer produced, Pol:

$$\text{Pol} = \int_{t_0}^{t_f} r_p V M_e\ dt = \frac{UA}{-\Delta H} \int_{t_0}^{t_f} (T - T_j)\ dt = K_{cal} \int_{t_0}^{t_f} (T - T_j)dt \tag{5}$$

where the constant parameters U, A and ΔH were collected in a single calibration constant K_{cal}.

In this way, assuming ΔH constant, the calibration constant value at the end of the test can be known, and Equation (4) enables the calculation of the instantaneous polymerization rate obtained along the run. The instantaneous catalyst activity is given by:

$$A = \frac{r_p M_e}{[C]} \tag{6}$$

where A is the instantaneous catalyst activity, gr/mol-s; [C] is the catalyst concentration, mol/L.

r_p determination by ethylene flow measurement

The ethylene is fed from a higher pressure reservoir (also constant).The instantaneous polymerization rate is given by a quasi-steady state mass balance for the ethylene where the accumulation term has been neglected, resulting in:

$$r_p = \frac{M_e}{V} F_e \tag{7}$$

As in the case of the calorimetric method, the relationship between catalyst activity and polymerization rate is obtained with Equation (6).

Results of the kinetic monitoring realized in two polymerization tests with the catalytic system S_1 are shown in figure 2.

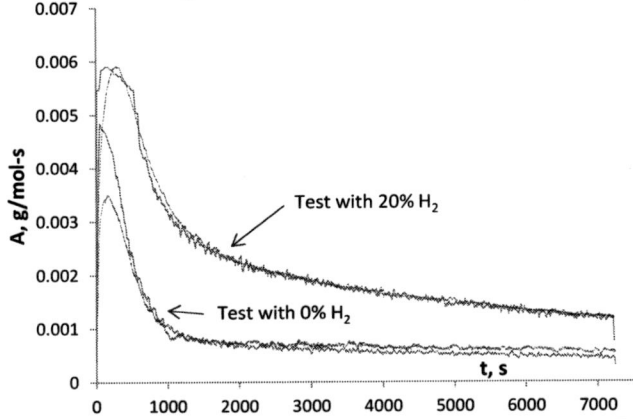

Figure 2. Comparison of calorimetry (—) and direct method (ethylene flow measurement) (- - -) in the monitoring of the catalytic activity, A, of system S_1.

For the experiments using hydrogen, a mixture of hydrogen with ethylene with partial pressures corresponding to H_2/C_2 of 50 psig/1000psig was previously prepared in the ethylene cylinder. The kinetic behavior of the system in solution or homogeneous phase showed a high initial activation rate of the catalyst, exhibiting the highest rate of monomer consumption during the first hour of reaction, and then decaying in the second hour, suggesting the deactivation step of the catalyst.

CONCLUSIONS

It can be stated that calorimetry and direct measuring of ethylene flow are methods conveying essentially the same kind of information about the instantaneous activity of Ziegler-Natta catalysts. The results presented in this study support the intuitive idea that in the isoperibolic operation, the reaction temperature excursion is an indicator of the activity of the catalyst. Also, this work presents enough information to guide the implementation of the techniques in a reactor laboratory, in order to follow online the instantaneous activity of Ziegler-Natta catalysts.

REFERENCES

1. J.B.P Soares, T. McKenna, C.P. Cheng, In *Polyolefin Reaction Engineering*, J.M. Asua Ed.; Blackwell Publishing, Chapter **2**, pp 29-117 (2007).
2. I. Kouzai, B. Liu, T. Wada, M. Terano , *Macromolecular Reaction Engineering* **1**, 160-164 (2007).
3. H. Kiyoung-Su, Y. Kee-Yon, R. Hyun-Ku, *Journal of Applied Polymer Science* **79**, 2480-2493 (2001).
4. F. Rincon, M. Esposito, P. Araujo, C. Sayer, A. Le Roux , *Macromolecular Reaction Engineering* **7**(1), 24-35 (2013).
5. M. Esposito, C. Sayer, R. Machado, P. Araujo, *Macromolecular symposia* **271**, 38-47 (2008).
6. F. Korber, K. Hauschild, G. Fink, *Macromolecular Chemistry and Physics* **202** (17), 3329-3333 (2001).
7. F. Korber, K. Hauschild, M. Winter, G. Fink, *Macromolecular Chemistry and Physics* **202**(17), 3323-3328 (2001)
8. I. Altarawneh, V. Gomes, S. Mourtada, *Polymer International* **58**, 1427-1434 (2009).
9. V. F. Isse, N. Sheibat-Othman, T. F. L. McKenna, *The Canadian Journal of Chemical Engineering* **88**, 783-792 (2010).
10. H. U. Moritz, In *Polymer Reaction Engineering*, Reichert and Geisler Eds., Weinheim, Germany Verlag-Chemie, pp 248-266 (1989).

Mater. Res. Soc. Symp. Proc. Vol. 1613 © 2014 Materials Research Society
DOI: 10.1557/opl.2014.166

Friction and wear behavior of a PMMA-SiO$_2$ coating on hardened steel

Luis E. Torres-Parga, Carolina Hernández-Navarro, Karla J. Moreno-Bello, J.S. García-Miranda, Luis D. Aguilera-Camacho, Raúl Lesso-Arroyo, Benjamín Arroyo-Ramírez and Álvaro Sánchez-Rodríguez

Department of Mechanical Engineering, Technological Institute of Celaya, Celaya, Guanajuato, Mexico

ABSTRACT

Sol-gel coatings show an excellent chemical stability, oxidation control and enhanced corrosion resistance for metal substrates. An organic-inorganic hybrid consisting of poly (methyl methacrylate) (PMMA) and silica (SiO$_2$) was successfully synthesized in the form of solution, by using 3-(trimethoxysilyl) propyl methacrylate (TMSPM) as a coupling agent and cohydrolyzed with tetraethyl orthosilicate (TEOS) to afford chemical bondings to the forming silica networks by a sol-gel method. The as-synthesized hybrid material was subsequently characterized by Fourier Transformation infrared (FTIR) spectroscopy. PMMA-SiO$_2$ was applied as a protective film on hardness steel substrates by dip-coating. The thickness of the coating was 25 µm, while the roughness $R_a = 0.6$ µm. The wear and friction behavior of the coating on hardened steel (HS) was evaluated by a ball-on-disk test in dry conditions with a AISI steel ball as counterface applying 2, 4, 6, 8 and 10 N normal loads. Friction coefficient values (μ_k) were in the range of 0.76 to 0.99, whereas the lowest wear rate (k) was observed at 6N with a value of 1.30×10^{-4} (mm^3(Nm)$^{-1}$).

INTRODUCTION

Hardened steels (HS) are widely used in the automotive, gear, bearing, tools and die industry [1]. Sol–gel coatings improve the chemical and physical properties of the metal surfaces relative to corrosion, friction and wear without altering the original properties of strength and toughness of the substrate [2]. Further, the sol–gel method is an environmentally friendly technique of surface protection and had showed the potential for the replacement of toxic pretreatments and coatings, which have traditionally been used for increasing corrosion resistance of metals [3]. Recent work on PMMA/SiO$_2$ hybrids has been reported on the synthesis by a sol-gel procedure [4-7], as an anticorrosively coating on steel substrates [2,4] and the study of mechanical properties [6,8-10], among others. Tribological behavior of hybrid materials prepared by the sol-gel method and deposited by dip-coating had been reported for hybrid thin films on glass slides [5,11] and acrylic substrates [6,11], however, less information is available on their tribological properties as coating on metal, specifically on hardened steels (HS) substrates. Protective coatings for wear and corrosion protection are of interest to induce significant changes in the performance of HS applied as engineering components. Moreover, PMMA-SiO$_2$ coatings are flexible, a property conferred by the organic component, and thus they can undergo higher deformations before cracking if applied on steel substrates. In this work, PMMA/SiO$_2$ hybrid material was simple synthesized in the form of solution using a sol-gel approach based on the polymerization of

TEOS in the presence of PMMA and applied as coating on HS substrates. Ball-on-disk tests in dry sliding conditions were performed. The friction coefficient was obtained and wear rates were calculated. The main wear mechanism for the specimen was established after observation of worn tracks.

EXPERIMENTAL DETAILS

Tetraethyl orthosilicate (TEOS, 99%, Aldrich), methyl methacrylate (MMA, 99%, Aldrich), benzoyl peroxide (BPO, Aldrich), 3-(trimethoxysilyl) propyl methacrylate (TMSPM, 99%, Aldrich), hydrochloric acid, (HCl, Aldrich), ethyl alcohol was used as the organic solvent and distilled water was used to hydrolyze the organic inorganic precursor. The PMMA-SiO_2 hybrid material was synthesized as solution using a modified sol-gel technique [4]. First, PMMA was synthesized by the polymerization of 191.76 mmols of MMA by a free radical polymerization started using 0.386 mmols of BPO as the initiator. TMSPM solution was prepared (TMSPM:H_2O:EtOH=0.025:0.15:0.025 M) and stirred during 60 min. TEOS solution was prepared (TEOS:EtOH:H_2O=0.03:0.70:0.2) and a drop of HCL was used as catalyst, the solution was then stirred during 30 min. PMMA-SiO_2 hybrid material was prepared by the subsequent addition of the solutions previously prepared in the appropriate volume relation (PMMA:TMSPM:TEOS=0.1911: 0.1637:0.6371), then mixed during 30 min. The functional groups of the hybrid material were analyzed by Fourier Transformation infrared (FTIR) spectroscopy using a Perkin Elmer Spectrometer 400 with an Attenuated Total Reflectance (ATR) coupled. HS steel pieces (1.3 C, $\leq$0.4 Si, $\leq$0.4 Mn, $\leq$0.03 P, $\leq$0.02, 4.3 Cr. 4.8 Mo, 5.4 W, 4.1 V, bal.Fe, in wt.%), with a ϕ 25 mm and thickness of 7 mm and a surface roughness of R_a $\leq$ 0.4 μm, were applied as substrates. The coatings were prepared by dip coating of HS substrates in the PMMA/SiO_2 solution, with an immersion speed of 70 mm/min during 5 min. After the immersion process was completed, the samples were left in dry air for 3 min and then dried at 80°C during 4 h. The coating thickness was calculated using image analysis software in a Zeiss Imager A1m microscope. For this purpose, the cross-section was prepared by embedding the samples in epoxy resin, and polishing by conventional metallographic techniques. The average surface roughness (R_a) of the coatings and the thickness were analyzed using a MITUTOYO Surftest 402. Twenty measurements were made at the direction perpendicular to the coating surface in an area of 40 cm^2. Wear tests were carried out by a ball-on-disk method in dry on a CSM Instruments Tribometer. Kinetic friction coefficients (μ_k) were obtained directly of the Tribox 4.1 software. AISI steel 52100 ball, with a diameter of 6 mm with a hardness of 63 HRC, was previously cleaned with EtOH and slided on the HS substrate coated with the PMMA-SiO_2 hybrid material. The standard contact loads used were 2, 4, 6, 8 and 10 N. The sliding cycles and speed were settled at 10000 cycles and 0.10 m/s, respectively. The environment temperature during the test was maintained at 26 ± 1 °C with a relative humidity of 30–40%. Since the wear mass loss values of the samples were inconsistent and the difference in weight loss was negligible, we rather determined the volume loss (V) of the coating by a standard test method as indicated in the ASTM G99-05 [12], assuming that there is no significant pin wear. Whereas, wear rate (k), was calculated from the relationship given elsewhere [13]. After tests, widths of wear tracks were determined and worn surfaces were analyzed by an optical microscope; wear mechanisms were identified.

DISCUSSION

The obtained PMMA-SiO₂ coating was homogeneous with a thickness of 25μm and a roughness of $R_a = 0.6$ μm. A thin films composite of PMMA/SiO₂ /TiO₂ on microscope glad slides reported a thickness of 10 nm [5]. Whereas, sol-gel coatings for chemical protection of stainless steel show thicknesses of about 0.3 μm to 0.6 μm [2], however, sol-gel films were prepared by dip-coating using sonocatalyzed sols of a single layer deposition, also parameters of immersion and thermal treatments were different. Solukhin et al. [10] reported thickness in a range of 15-70 μm for hybrid cross-linked coatings consisting of polymer methacrylate matrices with dispersed nano-sized silica particles, applied on polycarbonate substrates by means of a Doctor Blades applicator. Figure 1 shows the presence of the characteristics PMMA and SiO₂ absorption bands in the PMMA-SiO₂ spectrum, showing the hybrid nature of the synthesized material from the sol-gel process. The strongest absorption band detected between 1300 and 700 cm⁻¹ is associated mainly with the inorganic contribution. Absorption peaks corresponding to Si-O-Si symmetric and asymmetric mode are identified at 816 cm⁻¹and 1046 cm⁻¹[4]. The wide band centered at 3342 cm⁻¹ is due to O-H stretching modes [9], which are mainly due to Si-OH groups, but can also be a result of contributions from residual ethyl alcohol and molecular water adsorbed or from the hydrolysis of TEOS and TMSPM in the sol-gel process [4]. The absorption bands detected in the range from 1293-1452 cm⁻¹ and 2888-2973 cm⁻¹ are related with C-H contributions [5]. Important organic contributions are detected at 1636 cm⁻¹ and 1711 cm⁻¹, which correspond to C-C and C-O, respectively [9].

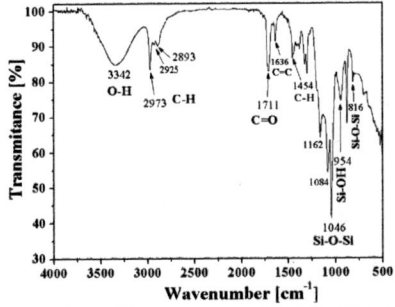

Figure 1. FTIR spectrum of PMMA-SiO₂ hybrid material.

Figure 2 shows the variation of friction coefficient (μ$_k$) as a function of the sliding cycles for PMMA/SiO₂ hybrid coating on HS substrates for each normal load applied. For all coatings, there is an initial range between 1677 cycles where can be noticed a change in the friction coefficient behavior. PMMA/SiO₂ coating tested at 6 N, presented a several fluctuation motion during the whole test. An increment in the friction coefficient could be noticed for coatings tested at normal loads of 8 N and 10 N. This effect could be attributed to the removing in each case of material debris by plowing or peel-off. For coatings tested at 4 N and 2 N applied loads no fluctuation was observed and a constant behavior was observed. The lowest mean friction coefficient was μ$_k$ = 0.76, obtained with a normal load of 4 N, while the highest mean value was

μ_k = 0.99, obtained with a normal load 8 N and the highest standard deviation value reported. The standard deviation of each one of the plotted values varied between μ_k = 0.09 to μ_k = 0.28. G. Gu et al. [5] evaluate the tribological properties of organic-inorganic hybrid (PMMA/SiO$_2$/TiO$_2$) thin films on glass substrates, proposed for protective lubricating. The friction test to coatings were performed by a reciprocating friction test with a sliding velocity (0.0015 m/s), a normal force (3 N) and just 8000 sliding cycles. They exhibit a low friction coefficient (0.09-0.11), these low values of friction coefficient are as lows as those reported by Alvarado-Rivera et al. [6] for SiO$_2$-PMMA hybrid films (0.1-0.15). Both values resulted even lower than those obtained for the present study (0.76-0.99), we can relate it to differences on the substrates; both studies on composite thin films were performed on microscope glad slides and acrylic substrates, respectively. Wear behavior of SiO$_2$-PMMA hybrid coatings reinforced with Al$_2$O$_3$ whiskers and nanoparticles [11], was studied on glass and acrylic substrates showing better performance on glass substrates but abrupt increments of the friction coefficient was also observed. The generated debris promotes an increase on the friction coefficient for these hybrid coatings as well as a variation due to the difference between substrates and reinforcement particles (0.2-0.75). The parameters applied for the studies mentioned before, are considerable lower than the ones applied for evaluate the PMMA-SiO$_2$ coating on HS substrates in this study. Table 1 presents the mean friction coefficients (μ_k) obtained for the PMMA-SiO$_2$ hybrid coating on HS substrates at different normal loads.

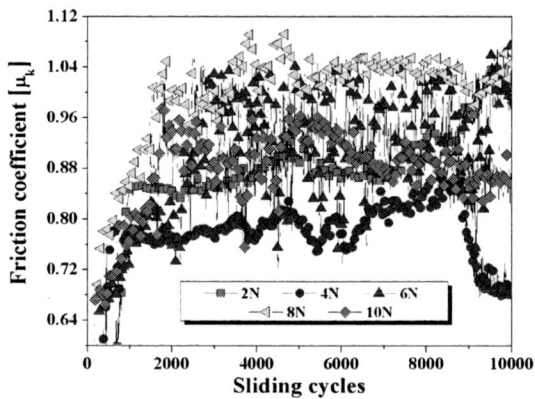

Figure 2. Friction coefficients of PMMA-SiO$_2$ hybrid material on HS substrates.

The lowest wear rate obtained in this study was k = 1.30 x10^{-5} mm^3/(Nm) at 6 N, while the highest value was k = 3.75 x10^{-5} mm^3/(Nm), observed at 2 N. The highest volume loss (V) was V= 0.55 mm^3, for 10 N, while the lowest was V = 0.07 mm^3 at 2 N. Table 1 shows the mean values for wear rate (k) and volume loss (V), it can be observed that wear rate (k) decreases as volume loss (V) increases. This behavior was also observed by Wang et al. [14] when PTE powder and nano-SiC particles were jointly incorporated into PEEK, the friction reduction and wear resistance capacities of the material became worse. This was due to the chemical reaction between nano-SiC particles and PTFE during the sliding process yielding SiF$_x$. This deteriorated the formation of the transfer film on the slider contact surface and ultimately resulted in poor tribological properties compared to binary PEEK/SiC composite and near PEEK [14].

Table I. Average values of: friction coefficient (μ_k), volume loss (V) and wear rate values (k), for the PMMA-SiO$_2$ hybrid coating on hardened steel substrates at different loads.

Load [N]	μ_k	V [mm^3]	k [x10^{-5}, mm^3(Nm)$^{-1}$]
2	0.85 ± 0.16	0.07	3.75
4	0.76 ± 0.10	0.16	1.90
6	0.88 ± 0.13	0.09	1.30
8	0.99 ± 0.28	0.28	1.74
10	0.88 ± 0.09	0.55	1.75

Standard deviations for volume loss and wear rate values are negligible.

Figure 3 shows the optical micrographs for worn surfaces at 2 N (a), 4 N (b), 6 N (c) and 10 N (d) normal loads. It could be notice that coatings subjected at normal loads of 2 N and 4 N are covered with a characteristic red-brown transfer layer. This could be related to the forces associated with adhesive and deformation process. Adhesive forces are often related to the shear strength of junctions made between contacting asperities of surfaces in contact with each other but in relative motion. The presences of oxide films on these asperities could clearly influence the resulting forces opposing motion and surface temperatures generated during contact [15]. This is not noticed at 6 N and 10 N, probably debris might function as solid lubricant. During sliding against a rough metal counterface, the metal asperities penetrate the softer polymer, and wear occurs via plastic deformation leading to shear, or micro-cutting. It could be observed that the predominant wear mechanism is abrasion wear, most common type of wear particularly in polymer composites. For polymer composites, many studies have shown that there are no fixed correlations of wear with mechanical properties. It is found that the extent of counterface modification or damage plays an equally important role [15].

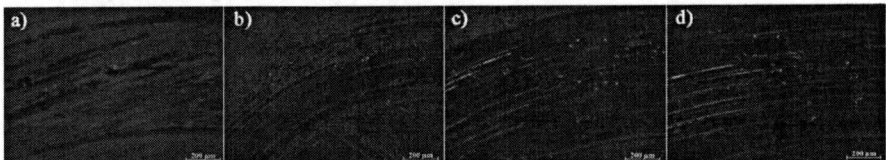

Figure 3. Optical micrographs of worn surfaces for PMMA-SiO$_2$ coating on HS substrates against steel ball at 2 N (a), 4 N b), 6 N (c) and 10 N (d), at 10x.

CONCLUSIONS

In this work, a hybrid material based on a PMMA polymer matrix and a SiO$_2$ component (PMMA-SiO$_2$) was successfully prepared in the form of solution by a sol-gel method and deposited by dip coating technique on hardness steel (HS) substrates with the aim of providing protective and homogeneous properties to the surface. A homogeneous coating without cracks was obtained with a thickness and roughness of 25 µm and 0.6 µm, respectively. We have reported the wear and friction behavior of a PMMA-SiO$_2$ coating on HS substrates by the ball on disk method. The lowest mean friction coefficient was $\mu_k = 0.76$, applying a normal load of 4 N, while the highest mean value was $\mu_k = 0.99$ with 8 N. It was observed that wear rate (k) decreases as volume loss (V) increases. The lowest wear rate obtained in this study was k = 1.30 $\times 10^{-5}$ mm^3/(Nm) at 6 N, while the lowest volume loss was V=0.07 mm^3 at 2 N. The predominant wear mechanism was abrasion wear. PMMA-SiO$_2$ hybrid material could be applied as protective coating on HS surface to improve wear resistance, friction coefficient, and as a potential protection from corrosion in various mediums and practical applications.

ACKNOWLEDGMENTS

The authors gratefully acknowledge the financial support of CONACYT-CONCYTEG grant GTO-2011-C03-161566. Also we would like to express our gratitude to M.Sc. Victor Alfonso Morales Nieto for his help with the experimental procedures.

REFERENCES

1. J. F. Archard and W. Hirst, *Proc. R. Soc. Lond. A* **236**, 397 (1956).
2. P. de Lima, M. Atik, L. A. Avaca and M. A. Aegerter, *J. Sol-Gel Sc.Techn.* **2**, 529 (1994).
3. D. Wang and G. P. Bierwagen, *Prog.Org. Coat.* **64**, 327 (2009).
4. Y. Jui-Ming, W. Chang-Jian, L. Wen-Jia and M. Yi-Wen, *Surf. Coat. Technol.* **201**, 1788 (2006).
5. G. Gu, Z. Zhang and H. Danga, *Appl. Surf. Sci.* **221**, 129 (2004).
6. J. Alvarado-Rivera, J. Muñoz-Saldaña and R. Ramírez-Bon, *J. Sol-Gel Sci.Technol.* **54**, 312 (2010).
7. Y. Y. Yu, C. Y. Chen and W. C. Chen, *Polymer* **44**, 593 (2003).
8. C. A. Avila-Herrera, O. Gómez-Guzmán, J. L. Almaral-Sánchez, J. M. Yáñez-Limón, J. Muñoz-Saldaña and R. Ramírez-Bon, *J. Non-Cryst. Solids* **352**, 3561 (2006).
9. L. Kyu-Hyeon and R. Sang-Hoon, *Biomater* **30**, 3444 (2009).
10. V. A. Soloukhina, W. Posthumus, J. C. M. Brokken-Zijp, J. Loos and G. de With, *Polymer* **43**, 6169 (2002).
11. J. Alvarado-Rivera, J. Muñoz-Saldaña, A. Castro-Beltrán, J. M. Quintero-Armenta, J. Almaral-Sánchez and R. Ramírez-Bon, *Phys. Stat. Sol. C* **14**, 4254 (2007).
12. ASTM G99-05. Standard test method for wear testing with a pin-on disk apparatus, American Society for Testing and Materials (ASTM International), USA (2005).
13. R. Sheryl, D. A. Paul and A. P. Lisa, *J. Biomed. Mater. Res. A* **92**, 1500 (2009).
14. Q. H. Wang, Q. J. Xue, W. M. Liu and J. M. Chen, *Wear* **243**, 140 (2000).
15. J. Robert and K. Wood, *J. Phys. D: Appl. Phys.* **40**, 5502 (2007).

Mater. Res. Soc. Symp. Proc. Vol. 1613 © 2014 Materials Research Society
DOI: 10.1557/opl.2014.167

Enhanced Structural Behavior of Systems Polypropylene-Carbon Nanotubes in Acidic Medium

Felipe Avalos-Belmontes[1], Miguel Flores-Godina[1], RosaNarro-Cespedes[1], Adali Castañeda-Facio[1], Martha Castañeda-Flores[1], Maura Tellez-Rosas[1], Luis Ramos-deValle[2], and Roberto Zitzumbo-Guzman[3]

[1] Universidad Autónoma de Coahuila, V. Carranza sn, 25000, Saltillo Coahuila, México
[2] Centro de Investigación en Química Aplicada, E Reyna sn, 25204, Saltillo Coahuila, México
[3] Centro de Innovación Aplicada en Tecnologías Competitivas, México.

ABSTRACT

The effect of carbon nanotubes (CNTs) on the thermal and chemical stability of polypropylene (PP) when subjected to oxidation in a fuming nitric was evaluated. The effect of CNTs on the crystalline morphology and melting and crystallization temperature of PP was studied. The results shown a thermal stability increased markedly; the decomposition temperature, increased from 293°C for pure PP to   320°C for PP with CNTs. The crystallization temperature increased perceptibly in presence of CNTs. The oxidative degradation with nitric acid produced a reduction in molecular weight; however, this negative effect was less pronounced in the PP compositions with carbon nanoparticles. The morphological changes evaluated with X-ray diffraction showed that the alpha type crystallinity remains, irrespective of the nucleating agent, and the intensity ratios between reflections peaks was taken as an indication of an increasing nucleating efficiency.

INTRODUCTION

With the addition of filler particles, and especially nanoparticles, the polypropylene can be tailored to cover many engineering applications. There are many studies that report on the use of different nanoparticles in polypropylene in order to impart or modify a given property. Inorganic and organic [1-4] nanoparticles have been used to enhance the mechanical, thermal, flame resistance and barrier properties of PP. In most cases, when added into PP, these nanoparticles act as nucleating agents, promoting thus the heterogeneous crystalline growth.

In a recent work [4], the authors reported on the excellent nucleating ability of carbon nanoparticles (conc. 0.05%), as compared to commercial nucleating agents for PP. This excellent nucleating capacity of CNT and CNF has also been reported by other authors [5-6], though some authors have found that above 4 wt% this ability reaches a plateau and remains constant.

On another hand, an important application of PP is in lead-acid car battery cases in which their outstanding properties make it the ideal option for long time performance. In these applications the PP is in contact with highly corrosive substances which can alter its composition and structure and significantly reduce its chemical and mechanical properties.

The purpose of the present work is to study the relationship between the increase in the thermal and chemical resistance and the enhanced nucleation induced by carbon nanotubes. The reason for the use of very low nucleating agent concentrations (0.05 wt%) is to avoid the excess of nucleating particles from obstructing the spherulitic growth, as reported by other authors when using higher concentrations of carbon nanoparticles.

EXPERIMENTAL DETAILS

The PP used in this study was a commercial grade isotactic from Total Petrochemicals, USA, with a MFR of 15 g/10 min. The CNTs used in this study, were multi-walled carbon nanotubes, with surface area of 200 m^2/g, from Nano-Lab, Inc., USA.

A series of PP compositions with 0.05 wt% of CNT were prepared via melt mixing for 10 min, at 170 °C and 50 rpm, in a Brabender plastometer. Plain PP was also passed through the mixer to produce blank samples. The compositions were then compression molded to produce 20x20x0.3 cm laminates, from which, 10 cm long and 1 cm wide strips were cut. The strips were immersed in fuming nitric acid at 80 °C for periods of 2, 8 and 21 h, after which, the samples were washed with distilled water until neutral pH, and finally, thoroughly washed with acetone.

A Perkin-Elmer DSC-7 was used, in order to establish the changes of the thermodynamic variables. Each sample was first heated from room temperature to 180 °C, held there for 2 min and then cooled down to room temperature at 10 °C/min. XRD analysis was performed in a Siemens D5000 (25 mA, 35 kV) using CuKa X-ray radiation, at 0.6 degrees/min from 1 to 15 degrees. For thermal degradation a Shimadzu TGA-50 analyzer was used, from RT to 400 °C, at 10 °C/min, under air atmosphere. And for molecular weight determination a chromatograph GPC V-2000, with a refraction index detector and "styragel" columns, was used.

DISCUSSION

Figure 1 shows the TGA results of the pure PP and the PP with 0.05 wt% of CNF. Samples were not immersed in nitric acid. It can be observed that up to ~220 °C, the samples remain unaltered; but it is at this temperature where the pure PP presents its onset of decomposition; the other sample presents it at a higher temperature. Also, it can clearly be observed that a 10 wt% weight loss, as well as a 50 wt% loss, occurs first for the pure PP, at 262 and 293 °C, respectively; with the carbon nanoparticle compositions producing the highest decomposition temperatures, as shown in the inserted table.
These results corroborate the enhanced thermal and chemical stability of the nucleated compositions. This enhanced stability is related to the crystalline structures produced in PP, in the presence of nucleating agents.

The more marked effect on the increase in the decomposition temperature of the carbon-nucleated compositions may be due to the barrier effect of the carbon nanoparticles, which when well dispersed, form a barrier that obstructs the oxygen diffusion [9], retarding thus the thermo-oxidative degradation of the polymer.

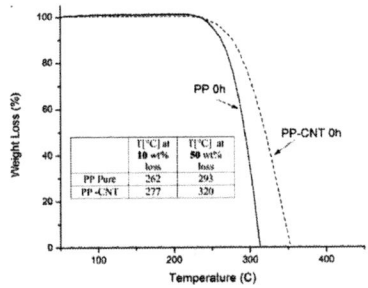

Figure 1. TGA curves in an air atmosphere, for pure PP and PP-CNT (0.05 wt%).

On another hand, the Figure 2 shows that after 8 hours in nitric acid, the molecular weight, as taken at the peak of the GPC curve, decreased from 141,000, for the original PP, to 68,000 (PP pure), and 79,500 (PP-CNT). The differences in molecular weight after the 8 hours in nitric acid are noticeable, though small. This is assumed to be because the nucleated crystalline structures provide a slightly better chemical stability for the PP nanocompounds. In addition, the carbon nanoparticles tend to act as free radical scavengers, as reported elsewhere [10], further retarding a little bit the PP degradation reaction.

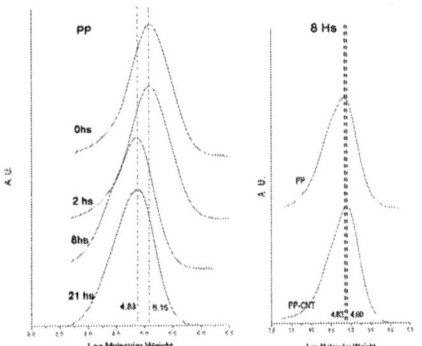

Figure 2. Variation of molecular weight with the time of immersion in nitric acid for pure PP, and nucleated PP, after 8 hours.

Figure 3 shows the TGA curves of the pure PP and nucleated PP, after 8 hours of immersion in nitric acid. First, it can be observed that the samples remain unaltered up to ~240 °C, that is, 20 °C higher than the results obtained for the same samples, but without immersion in nitric acid.

Secondly, it can be observed that after 8 hours in nitric acid, a 10 as well as a 50 wt% loss occurs first in the pure PP, followed by the nucleated composition PP-CNT, attaining the highest decomposition temperatures, as shown in the table in the insert.

Finally, comparing the results in Figures 3 versus 1, it can be observed that the samples immersed in nitric acid in Figure 3, present a weight loss of 10 and 50 %, at temperatures ~40 and ~70 °C higher, respectively, than the corresponding temperatures for the samples without immersion in nitric acid in Figure 1.

Considering that the immersion in nitric acid tends to wash-out the amorphous fraction of the PP samples, then, the immersed samples surely have a greater crystalline fraction than the non-immersed samples, producing thus the marked differences in thermal stability, which could be attributed entirely to the greater crystalline fraction. In addition, the smaller differences between the decomposition temperatures of the pure and nucleated PP in Figure 4 can be attributed to the use of a nucleating additive, CNT.

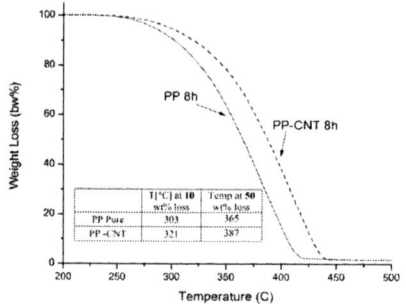

Figure 3. Weight Loss curves of pure PP and the PP- CNT nucleated, after immersion in nitric acid for 8 h.

The XRD diffractograms in Figure 4, show the position of the crystalline planes of pure PP and of the nucleated PP, without immersion in nitric acid and with 8 hours in nitric acid. It can be observed that the nanoparticles have no effect on the position of the diffraction signals.

This indicates that the crystalline type does not change due to the presence of the nanoparticles, but remains the same alpha crystals. Both, pure PP and PP/Nanoparticles present the same peaks at 2(theta) = 13.9, 16.7, 18.3, 21.6 and 21.9 degrees, which correspond to the planes (110), (040), (130), (111) y (041), respectively, of this alpha morphology.

In addition, a clear tendency for the increase of the intensity ratio between the (040) (16.7 degrees) and the (110) (13.9 degrees) reflections can be observed. This increase can be taken as an indication of an increasing nucleating efficiency, as reported elsewhere [4].

118

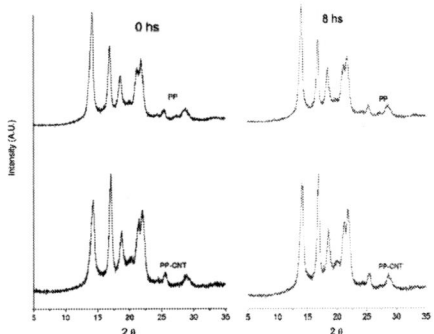

Figure 4. XRD diffractograms of pure PP and nucleated PP, without treatment in nitric acid and after 8 hours in nitric acid.

The fusion thermograms shown that the peak fusion temperatures are very similar between that for pure PP, 150.6 °C and that for the nucleated PP compositions, 151.5 °C. However, the fusion interval appears to be narrower for the PP compositions with nucleating agent, surely due to the greater crystal size homogeneity (Figure 5).

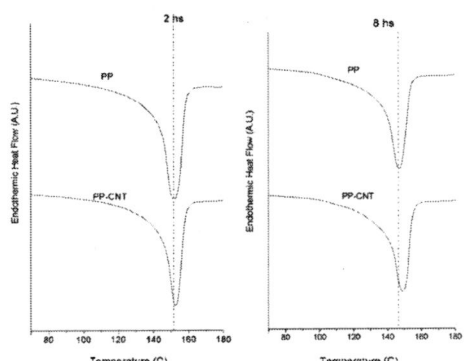

Figure 5. DSC traces of pure PP and nucleated PP, without treatment in nitric acid and after 8 hours in nitric acid.

On another hand, the Tm tends to show two distinct values; a higher value at the beginning and after 2 hours in nitric acid and a lower value after 8 and 21 hour in nitric acid. Nonetheless, Tm increases very slightly with the immersion time in nitric acid, from zero to 2 hours, after which,

Tm decreases. This slight increase could be attributed to the rearrangement of the non integer folded chains into the crystalline lattice, which increase the crystal size [11] and not because of any change in molecular weight; for according to the GPC results, any variation in molecular weight appears after 8 hours in nitric acid.

However, after 8 hours in nitric acid, Tm decreases, which may be assumed to be due to the formation of smaller crystalline structures, which may in turn be attributed to the presence of NO2 groups at the chain ends.

The crystallization temperature on the other hand, increased from 107 °C for the pure PP to 114 °C, for PP with CNT, respectively, which could be assumed to be due to the greater nominal surface area of the CNT.

CONCLUSIONS

Considering that the amount and morphology of the crystalline structure in a semicrystalline polymer greatly affects in a positive way its chemical and mechanical properties, it can be concluded that the inclusion of carbon nanoparticles as nucleating agents in PP would ensure better thermal and chemical resistance when subjected to aggressive environments.

The oxidative degradation with nitric acid produced a reduction in molecular weight, however, this was much less pronounced in the PP compositions with carbon nanoparticles.

CNT nanoparticles, increase markedly the thermal and chemical properties of PP, acting as very efficient nucleating agents. And the decomposition temperature increased from 293 °C for the pure PP to 320 °C, for the PP compositions with CNF.

REFERENCES

1. B. Kandola, G. Smart, A. R. Horrocks, P. Joseph, S. Zhang, and T. R. Hull, *J. Appl. Polym. Sci.* **108**, 816 (2008).
2. T. Zaharescu, S. Jipa, W. Kappel, and P. Supaphol, *Macromol. Symp.* **242**, 319 (2006).
3. Q. Wei, L. Yu, R. Mather, and X. Wang, *J. Mater. Sci.* **42**, 8001 (2007).
4. S. Reyes-deVaaben, A. Aguilar, F. Avalos, and L. F. Ramos-de Valle, *J. Thermal Analysis and Calorimetry* **93**, 947 (2008).
5. A. Funck, and W. Kaminsky, *Composites Sci. and Technol.* **67**, 906 (2007).
6. Y. Peneva, M. Valcheva, L. Minkova, M. Miuk, and M. Omastov, *J. Macromol. Sci.* **B47**, 1197 (2008).
7. H. E. Miltner, N. Grossiord, K. Lu, J. Loos, C. E. Koning and B. Van Mele, *Macromolecules* **41**, 5753 (2008).
8. X. Chen, S. Wei, A. Yadav, R. Patil, J. Zhu, R. Ximenes, and L. Sun, Zh. Guo, *Macromol. Mater. Eng.* **296**, 434 (2011).
9. G. Gorrasi, M. Sarno, A. Di-Bartolomeo, D. Sannino, P. Ciambelli, and V. Vittoria, *J Polym. Sci. Part B: Polym Phys*, **45**, 597 (2007).
10. F. Avalos, I. Zapata, L. F. Ramos-deValle, R. Zitzumbo, S. Alonso S., *J. Polym. Sci. Part B: Polym. Phys.* **47**, 1906 (2009).
11. B. K. Satapathy, M. Gans, R. Weidisch, P. Potschke, D. Jehnichen, and T. Keller, *Macromol. Rapid Commun.* **28**, 834 (2007).

Mater. Res. Soc. Symp. Proc. Vol. 1613 © 2014 Materials Research Society
DOI: 10.1557/opl.2014.168

Electrical Behavior I-V Theoretical-Experimental OLEDS

José M. Burgoa[1], Cecilia González-Medina[1], Ramón Gómez-Aguilar[1] and Jaime Ortiz-López[2]

[1]Unidad Profesional Interdisciplinaria de Ingeniería y Tecnología Avanzadas-IPN, Av. Instituto Politécnico Nacional 2580, México City, 07540.

[2] Escuela Superior de Física y Matemáticas-IPN, Av. Instituto Politécnico Nacional Edificio 9, Unidad Profesional Adolfo López Mateos, Zacatenco, Mexico City, 07738.

ABSTRACT

We develop a program (within MATLAB software environment) to numerically simulate current-voltage characteristics of a bilayer organic light-emitting diode (OLED). The program is based on the Poole-Frenkel and Schottky continuous quantum models which take into account the geometry of thin films and their emission parameters in the calculation of charge carrier and current density in organic materials. Simulations are performed for OLEDs with A/EML/C and A/HIL/EML/C architectures where A=anode, HIL=hole injection layer, EML=emissive layer and C=cathode. For EML we assume MEH-PPV and MDMO-PPV derivatives of poly-para-phenylene-vinylene (PPV) polymer semiconductor, and for HIL we use PEDOT:PSS. The results of simulation are compared with experimental results obtained from actual OLED devices constructed in our laboratory. For comparison we also use the commercial software SimOLED to simulate the devices under similar architectures. We find in general a fair agreement between the simulated and measured behavior except for a few orders of magnitude difference in the current.

INTRODUCTION

From a theoretical point of view, understanding of the behavior of organic light-emitting diodes with geometry ITO/MEH-PPV/Metal and ITO/PEDOT:PSS/MEH-PPV/Metal is possible through analysis based on the Schottky diffusion model and the Poole-Frenkel mobility model [1,2]. Theoretical modeling of the electrical response of OLEDs requires four essential parameters: film thickness of the organic semiconductor, work function of contacts, temperature of operation, and effective area of the electrodes. In this work we use the SimOLED (Sim4TEC) commercial software and our own MATLAB programs to model the behavior of OLEDs constructed in our laboratory. Since for the construction of OLEDs intervene parameters such as thickness, conductivity of the organic semiconductor and interface effects between the organic material and the contacts we consider eddy current models that provide a closer description of the behavior observed in actual physical devices. Typically, an OLED is comprised of: (a) a transparent cathode made of a material with high work function, usually indium-tin oxide (ITO); (b) an organic semiconductor active layer, usually deposited by spin casting this material dissolved in a volatile solvent, (c) an anode made of a metal or alloy (such as GaIn eutetic) of low work function. When excitons decay radiatively, the generated light emerges through the transparent region of the device. For conjugated polymers, singlet excitons are the only ones that decay radiatively.

EXPERIMENTAL DETAILS

The samples studied were prepared from luminescent polymer powders MDMO-PPV and MEH-PPV which were obtained from Sigma Aldrich. Thin films of these polymers were prepared by spin-coating starting from 0.5 wt% solutions of both polymers in tetrahydrofuran (THF) and in chloroform (CF) [3]. The films were deposited on fused quartz and on ITO-covered glass substrates (1 in square). The substrates are subjected to a process of ultrasonic cleaning in deionized water, acetone, acetonitrile and isopropyl alcohol each for periods of 10 min with a bath temperature of 60 °C. After leaving the last bath the substrates were dried with nitrogen gas. Spin coating deposition was done between 500 and 800 rpm for 30 or 40 seconds, finally using a PET-ITO mask for metal anode deposit GaIn.

THEORY

Methods based on quantum mechanics allow the study of organic semiconductors and OLED modeling by considering the device as a continuous system. The current density-voltage (J-V) characteristics of an OLED can be treated theoretically with Schottky model and the Poole-Frenkel equation of motion because the device becomes an electronic system that obeys the Pauli exclusion principle. Schottky model is used to represent the flow of current between the anode and the organic material [4]. If q is the elementary charge, μ the mobility [5], E the electric field, N the density of charge carriers, φ_t is the injection barrier height, γ is the barrier due to reduced electric field, ε_0 is the permittivity of vacuum, and ε_d is the insulator's dynamic permittivity. As the mobility in the semiconductor depends on the field, the expression adjusts to the model as a result of thermal excitation of electrons from the traps of the insulator aided by the applied electric field eq. 1.

$$J \cong \frac{E}{T}\exp\left\{-q\left[\varphi_t - \sqrt{\left(\frac{qE}{\pi\varepsilon_0\varepsilon_d}\right)}\right]\bigg/kT\right\} \qquad (1)$$

Characterizing the device by the metal insulator semiconductor model (MIS) allows to evaluate J as a function of the thickness (d) and the dynamic permittivity ε_d eq. 2:

$$J \cong \frac{V}{d}\exp\left\{-q\left[\varphi_t - \sqrt{\left(\frac{q(V/d)}{\pi\varepsilon_i}\right)}\right]\bigg/kT\right\} \qquad (2)$$

The contribution of eddy currents in a diode is given by the Richardson-Schottky parameter and time of flight (TOF) model. If N is the number of free charges on the material, e is the elementary charge, d is the film thickness, V the applied voltage eq. 3, and w study area eq. 5. Equation 4 shows the contribution of the device while displaying equation 6 to calculate the current device and considering saturation current and the term ohmic described in eq. 7.

$$I = \frac{Ne}{d} v \text{ ; ideal current in diode} \quad (3)$$

$$I_D = I_0 \, exp\left(\frac{qV}{nkT}\right) \text{ ; } device\text{contribution} \quad (4)$$

where

$$I_0 = AT^2 w \, exp\left(-\frac{q\varphi_t}{kT}\right) \text{ ; saturation current} \quad (5)$$

Finally:

$$i = I_D + I_\Omega \quad (6)$$

$$I_\Omega = wV exp(\sqrt{V}) \text{ ; eddy current contribution} \quad (7)$$

The equivalent circuit for an OLED consists of a contact resistance Rc in series with a parallel combination of a resistor Rp and a capacitor Cp. The contact resistance comes from the ITO anode and the contact between the electrodes and the organic layer. The parallel resistor and capacitor values represent the bulk organic layer and vary according to the applied potential. The initial part of the circuit consisting of a capacitor and resistor in parallel is called 'capacitive' and the current flowing through it is phase-shifted with respect to the applied voltage. However, if the frequency of the applied voltage is very large, the impedance of the capacitor becomes zero and the 'capacitive' circuit behaves as a resistive one. Under typical conditions of operation of an OLED, a square wave is applied at low frequency (<50 Hz) with an average minimum voltage Vrms equal to the gap of the organic material in order to initiate the mobility of charge carriers injected from the contacts.

DISCUSSION

Modeling of OLED devices using the described theoretical treatment successfully represents the J-V characteristics provided the device is regarded as a continuous medium. Application of the models is limited to single and double layer devices. Simulation of the behavior obtained Poole-Frenkel and Schottky, models programmed with MATLAB, and with SimOLED through of the extended Gaussian disorder model (EGDM), and are presented in Figures 1a, 1b, 2b, 3b and 2a, 3a respectively. The contribution of eddy currents in the behavior is not important in the case of MDMO-PPV but for MEH-PPV with 200 nm thickness a large deviation can be observed with respect to the ideal device (SimOLED), due to the field, carrier-concentration and temperature dependence of the mobility in the GDM using the 3D-ME approach, and provided a parameterization of the results. In the 3D-MC method, we evaluate-the full hopping model, treats including all Coulomb interactions to difference of Schottky and Poole-Frenkel models.

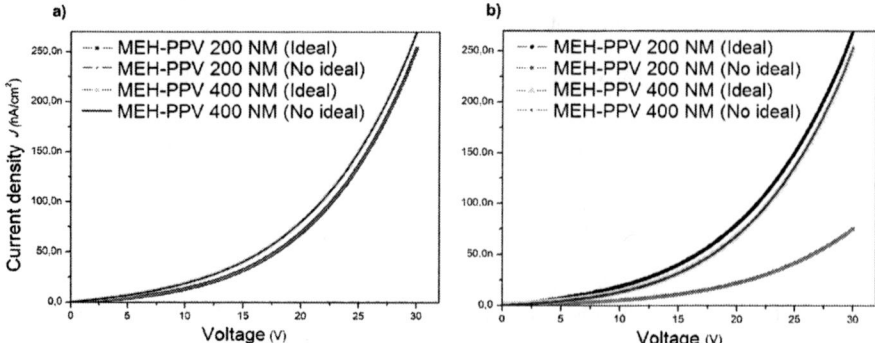

Figure 1. Simulation of J-V characteristics of OLED device with Schottky and Poole-Frenkel models programmed in MATLAB for MEH-PPV and MDMO-PPV with polymer film thicknesses of 200 and 400 nm.

The triple- and multi-layer structures that are simulated in SimOLED are not considered in our MATLAB program. However, the experimental characterization of the constructed OLEDs can be compared with the results for a simulated bilayer device programmed with MATLAB. Figures 2 and 3 show the J-V characteristic curves as follows: a) SimOLED geometry ITO (100 nm)/X:PPV(200 & 400 nm)/GaIn where X=MEH and MDMO; b) MATLAB, same geometry and same thicknesses as in a); c) experimental device geometry identical to the previous cases with a thickness of 120 nm semiconducting polymer [6]. We note that the simulated electrical behavior either by SimOLED or by MATLAB is linear between 8 and 12 volts but the corresponding current differs by two orders of magnitude. On the other hand if we study the behavior at voltages greater than 15 V, MATLAB simulation is in better agreement with the experimental behavior (see Figures 2, 3), however, the difference in current is still 3 orders of magnitude.

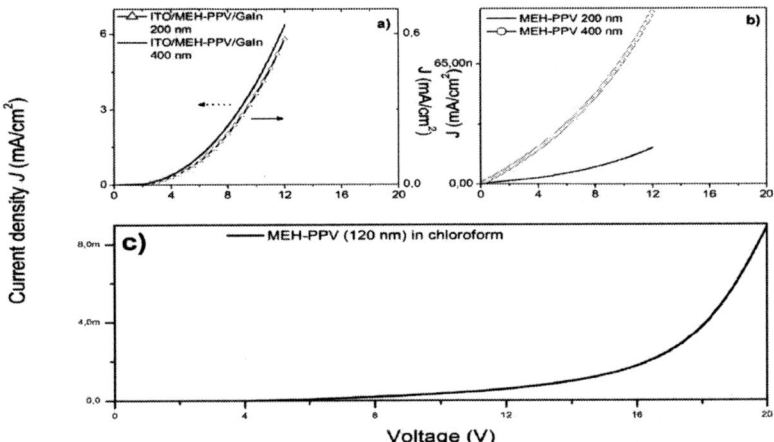

Figure 2. MEH-PPV OLED device characterization and simulation a) J-V curve by SimOLED, b) J-V curve by MATLAB implementation of Schottky y Poole-Frenkel models: c) Measured J-V characteristics of an actual OLED device.

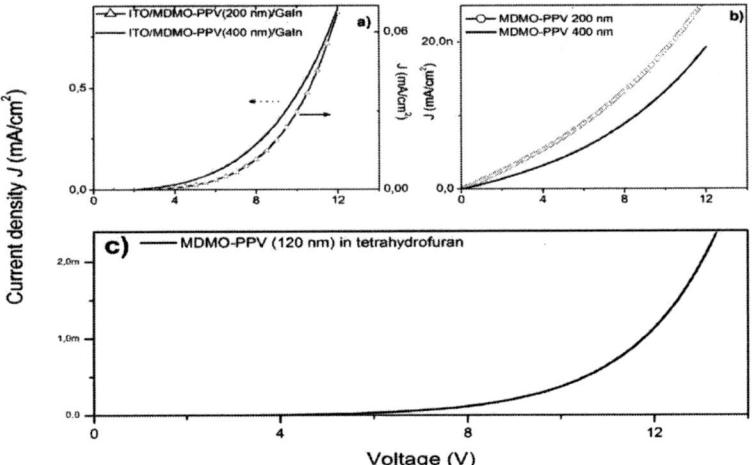

Figure 3. MDMO-PPV OLED device characterization and simulation a) J-V curve by SimOLED, b) J-V curve by MATLAB implementation of Schottky y Poole-Frenkel models: c) Measured J-V characteristics of an actual OLED device.

CONCLUSIONS

In this work we consider the construction and theoretical simulation of single and double layer OLED devices. Energy barriers to the injection of carriers in electrode interfaces, the interaction between materials, and the number of layers of the structure are just some of the important variables to be considered to obtain good performance, efficiency and extended lifetime in OLED devices [4,5]. Electrical characterization of OLED devices can be simulated with Schottky and Poole-Frenkel models [6], the resulting curves roughly described the behavior observed in actual OLED devices, which turn out to be quite helpful in deciding the best parameters in OLEDs device fabrication [7-10]. Both our experimental and theoretical results show an acceptable first approximation agreement of OLEDs device characteristics that help to determine the proper thickness of the emissive polymer layer, and to improve the performance and efficiency these devices [11,12].

ACKNOWLEGMENTS

Ramón Gómez-Aguilar acknowledges COFAA-IPN for a SIBE scholarship. Jaime Ortiz-López acknowledges COFAA-IPN for a SIBE scholarship and SIP-IPN for partial support through project number SIP-20130427.

REFERENCES

1. W. Brutting, S. Berleb, A.G. Muckl, *Synthetic Metals* **122,** 99 (2001).
2. T. Chu, O. Song, *Appl. Phys. Lett.* **90,** 203512 (2007).
3. R. Chang, J.H. Hsu, *Chem. Phys. Let.* **317**, 153 (2000).
4. A.L.Holt, J. M. Leger, S. A. Carter, *J. Phys. Chem.* **123**, 44704 (2005).
5. Gang Yu, Yongfang Li, Alan J. Hegger, *App. Phys. Lett* **73**, 111 (1998).
6. Y.Shi, J. Liu and Y. Yang, *J. Appl. Phys.* **87**, 4254 (2000).
7. W. F. Pasveer, J. Cottaar, C. Tanase, R. Coehoorn, *Phys. Rev. Lett.* **94**, 206601 (2005).
8. B. Tian, R. Schenk, *J. Phys. Chem.* **95**, 3191 (1991).
9. Shahul Hameed Ta, P.Predeep and M.R.Baij, *IJSSST* **11**, No. 4
10. L. D. Bozano, B. W. Kean, V. R. Deline, J. R. Salem, and J. C. Scott, *Appl. Phys. Lett.* **84**, 607 (2004).
11. Oleg Mitrofanov, Michael Manfra, *J. Appl. Phys.* **95**, 6414 (2004).
12. Middleman S & Hochberg AK. Process *Engineering Analysis in Semiconductor Device Fabrication.* McGraw-Hill. New York, USA. (1993) pp. 313.

Mater. Res. Soc. Symp. Proc. Vol. 1613 © 2014 Materials Research Society
DOI: 10.1557/opl.2014.169

Characterizing the Dosimetric Properties of MEH-PPV Using Thermoluminescence (TL)

Alejandro Ortiz-Morales[1,3], Ramón Gómez-Aguilar[1], Jaime Ortiz-Lopez[2], Epifanio Cruz-Zaragoza[3]

[1]Unidad Profesional Interdisciplinaria de Ingeniería y Tecnologías Avanzadas, IPN, Av. Instituto Politécnico Nacional 2580, Col. La Laguna Ticomán, 07340 México DF., México
[2]Escuela Superior de Física y Matemáticas, IPN, Av. Instituto Politécnico Nacional s/n, Col. San Pedro Zacatenco, 07738 México DF., México
[3]Instituto de Ciencias Nucleares, Universidad Nacional Autónoma de México, AP. 70-543, 04510 México DF., México

ABSTRACT

The thermoluminiscent properties of MEH-PPV and MDMO-PPV conjugated polymers were studied in order to verify if they are suitable for use as TL dosimeter. The dose response that was analyzed cover the wide dose range 0.34-5.44 kGy. The measured glow curves show complex structures which were evaluated with kinetic parameters based on the MO (Mix Order) model together with the CGCD (Computerized Glow Curve Deconvolution) homemade program which is useful to understand the mechanisms responsible for TL emission.

INTRODUCTION

The polymer poly(*p*-phenylene vinylene), abbreviated as PPV, consists of a chain of phenylene rings intercalated by vinylene bridges. MEH-PPV and MDMO-PPV are modified forms of the parent PPV polymer with additional methyl-based side groups attached to the polymer backbone. These polymers show optical absorption and photo- and electro-luminescence bands in the visible region which make them important candidates for many electronic applications. In the past twenty years, flexible electro-optical devices such as organic light emitting diodes (OLEDs), organic field effect transistors (OFETs) and organic photovoltaic cells (OPVCs) have been developed with the use of thin films of semiconducting conjugated polymers like MEH-PPV and MDMO-PPV [1-3].
Thermoluminescence(TL) technique has been widely used to know if some crystalline solids doped with different metallic ions[4][5] and rare earths[6] are candidates for Thermoluminiscent Dosimeters (TLD), such as: LiF:Mg:Ti(TLD-100)[7], CaSO₄:Dy(TLD-700)[8]. Recent studies of MEH-PPV have showed that this polymer could be used as a radiation sensor [9]. For this reason, interest has grown for the analysis of MEH-PPV and MDMO-PPV with TL.

The aim of this work, is to report if the MEH-PPV and MDMO-PPV polymers have suitable TL properties for their use in dosimetry.

EXPERIMENTAL DETAILS

Samples A and B used in this work were MEH-PPV and MDMO-PPV, respectively, obtained commercially from Sigma Aldrich. Details of the samples are given in Table 1, and the values of the atomic weight (Wt).

Table 1. Conditions of sample preparation.

Preparation	Sample	Mass [mg]	Molecular Weight (Mw) [mol]
A	MEH-PPV	4	150,000-250,000
B	MDMO-PPV	4	23,000

Samples (A, B) were deposited on aluminum disk in proper amounts and were exposed to gamma photons from ^{60}Co of irradiator GammaBeam 650PT of the ICN-UNAM with dose rate of 0.170 kGy/min. After irradiating at room temperature (RT), the samples were inserted in black bags to avoid any light effects and stored in a lead box to eliminate any possible contribution from environmental irradiation. All the samples were read out in only one session at the end of the experimental period. The TL reader system was a Harshaw TLD model 3500; a constant heating rate of 2.0 K/sec was used and nitrogen gas was allowed to flow into the reading chamber during the read out to eliminate any spurious signals. The TL emission was integrated from room temperature (300 K) up to 700 K.

DISCUSSION

Kinetic parameters

The kinetic parameters i.e., the activation energy (E) of the traps involved in TL emission, R_m and α have been obtained using CGCD program. The algorithm of the deconvolution program is based on the mix order equation [10],

$$I(T) = I_m \frac{\{\exp((1\square\square_m)/R_m)\square\square\}^2}{\exp((1\square\square_m)/R_m)} \frac{\exp\left\{\dfrac{E}{kT}\dfrac{T\square T_m}{T_m}\right\}\exp\left\{\dfrac{T^2}{T_m^2 R_m}\exp\left(\dfrac{E}{kT}\dfrac{T\square T_m}{T_m}\right)(1\square\square_m)\right\}}{\left\{\exp\left\{\dfrac{T^2}{T_m^2 R_m}\exp\left(\dfrac{E}{kT}\dfrac{T\square T_m}{T_m}\right)(1\square\square)\right\}\square\square\right\}^2}$$

(1)

where;
k is Boltzmann constant, I_m and T_m are the maximum luminescence intensity and temperature in the peak deconvoluted, other useful quantities are: $\Delta = 2kT/E$, $\Delta_m = 2kT_m/E$.
In the fitting procedure the goodness of fit was tested with the figure of merit (FOM) [11] given by:

$$FOM = \sum \frac{|Y_{Exp} - Y_{Fit}|}{A}$$

(2)

where Y_{Exp} , Y_{Fit} are the experimental and fitted data of the glow curve, respectively, and A is the integral of the fitted glow curve. A FOM equal or less than 5 % means a very good fit.

Molecular structure, glow curve and deconvolution

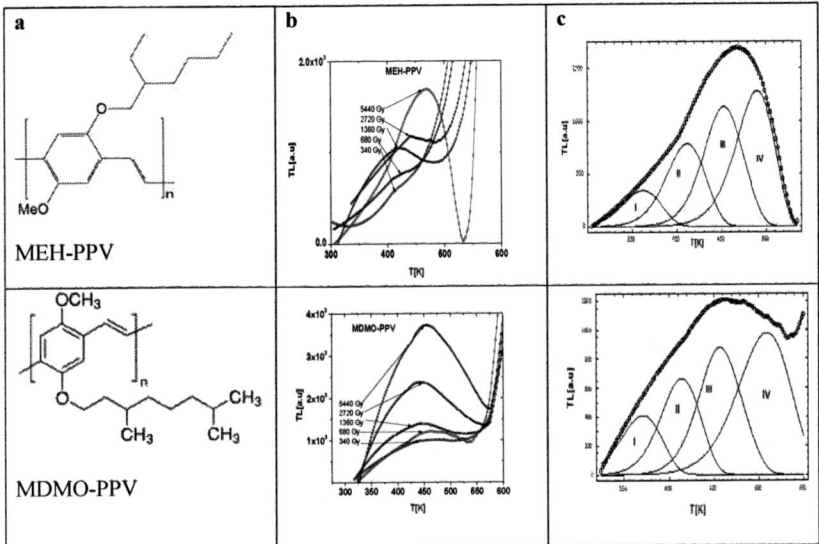

Figure 1. a) Molecular structure of MEH-PPV, and MDMO-PPV, **b)** Glow curves as a function of dose. The doses were 0.34, 0.68, 1.36, 2.72 and 5.44 kGy, and **c)**CGCD of the experimental glow curves (open circles) for a dose of 0.170 kGy at a dose rate of 0.170 kGy/min.

Glow curves (sample A, B) exhibit just a wide peak centered around at 450 K which grows as function of dose, induced for several trapping states.

TL signal as a function of dose

Two stages are seen in the TL response as a function of the dose for both preparations. At 340 Gy a monotonous growth in the TL signal is observed in MEH-PPV until 5.44 kGy (stage 2), while a contrasting behavior is displayed by MDMO-PPV with a decreasing trend until 2.72 kGy. Stage 1 displays a complex behavior in the TL signal for both samples, for MEH-PPV increases while for MDM-PPV decreases.

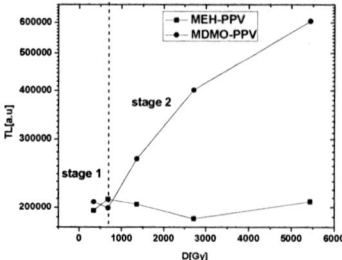

Figure 2. TL dose-response as a function of dose for MEH-PPV and MDMO-PPV.

Increasing dose seems to affect in different ways the response of MEH-PPV and MDMO-PPV. The effect of exposure to gamma radiation clearly breaks off bonds along the PPV backbone reducing therefore the conjugation length in both MEH-PPV and MDMO-PPV. At the same time, MEH and MDMO side groups may break as well and may also become detached from the PPV backbone. The different structure and configuration of the MEH and MDMO side groups may be the reason of the difference in the dose response of both polymers. In the disordered state in which both polymers are exposed to radiation, MEH and MDMO must acquire different positions relative to the PPV backbone. MDMO being larger than MEH could more efficiently wrap around the polymer chain and could act as some kind of shield against damage to the backbone. Therefore damage to MEH-PPV would be concentrated on the PPV chain while in MDMO-PPV extensive damage would be found mostly in the MDMO side chains. To check this interpretation it remains to apply alternative experimental techniques such as photoluminescence and Raman spectroscopies to samples before and after irradiation.

Table 2. Kinetic parameters as obtained by CGCD for samples (A, B).

Preparation A: MEH-PPV, FOM=0.04						Preparation B: MDMO-PPV, FOM=0.01					
Peak number	T_m[K]	R_m	□	E[eV]	s[s^{-1}]	Peak number	T_m[K]	R_m	□	E[eV]	s[s^{-1}]
I	370	1.09	0.46	0.60	1.87×10^7	I	370	1.71	0.43	0.59	1.62×10^7
II	420	1.16	0.51	0.78	3.26×10^8	II	420	1.41	0.54	0.80	7.45×10^8
III	470	1.05	0.6	1.0	1.06×10^{10}	III	480	1.0	0.73	0.97	5.08×10^9
IV	490	1.16	0.15	1.0	3.82×10^8	IV	500	2.0	0.31	0.82	1.40×10^7

CONCLUSIONS

MEH-PPV and MDMO-PPV glow curves show a complex structure and exhibit dependence on the irradiation dose. An intense glow peak (450 K) appears for samples A and B. The glow curves for the two samples have been deconvoluted with four peaks and the activation energy in the range 0.59-1.0 eV was calculated assuming mix order kinetics model. These energy values may be attributed to the binding energy values of the exciton, reported between 1 eV and kT to primary excitations, for instance, free charges (holes and electrons) [12]. Another way a theoretical estimate of the exciton binding energy in PPV based on effective mass calculations lead to the value of 0.4 eV [13], which coincides with the experimental values of films of PPV-derivatives [14].

ACKNOWLEDGMENTS

The authors are grateful to Benjamin Leal (ICN- UNAM) for their assistance in irradiation of samples. Ramón Gómez-Aguilar acknowledges COFAA-IPN for a SIBE scholarship. J. Ortiz-López acknowledges COFAA-IPN for a SIBE scholarship and SIP-IPN for partial support through project number SIP-20130427.

REFERENCES

1. M. T. Bernius, M. Inbasekaran, J. O'Brien, and W. Wu, Adv. Mater., 12, 1737 (2000).
2. S. Günes, H. Neugebauer, and N.S. Sariciftci, Chem. Rev., 107, 1324 (2007).
3. A. Kraft, A.C. Grimsdale, and A.B. Holmes, Angew. Chem. Int. Ed., 37, 402 (1998)
4. A. Ortiz-Morales, C. Furetta, G. Kitis, A. Negrón Mendoza and E. Cruz Zaragoza. Rad. Eff. Def. Sol. 161 (7), 383-393 (2006).
5. E. Cruz-Zaragoza, M. Barboza-Flores, V. Chernov, R. Meléndrez, B.S. Ramos, A. Negrón-Mendoza, J.M. Hernández, and H. Murrieta. Radiat. Prot. Dosim. 119 (1-4), 102-105 (2006).
6. España, T. Calderón, F. Cussó, F. Jaque, G. Lifante and P.D. Townsend. Nucl. Tracks Radiat. Meas. 20 (4), 605-607 (1992).
7. Y. Horowitz and D Yossian 1993 J. Phys. D: Appl. Phys. 26 1331
8. Azorin, J., Gonzalez, G., Gutierrez, A., Salvi, R., 1984. Health Phys. 46, 269–274.
9. Martins, D., Hempel, J.P., Andrade, A.M. IEEE Sensors Journal, Vol. 9, No. 7, July 2009.
10. Kitis, G., J.M.Gomez Ros., 2000. Nuclear Instruments and Methods in Physics Research A 440, 224-231.
11. Balian, H.G., Eddy, N.W., 1977. Nucl. Instr. and Meth. 145, 389-395.
12. Primary Photoexcitations in Conjugated Polymers: Molecular Exciton versus Semiconductor Band Model ed. By N. S. Sarixiftci (World Scientific Publishing, Singapure, 1997)
13. P. Gomes Da Costa, E. M. Conwell., Phys. Rev. B 48, 1993 (1993).
14. S. Barth, H. Bässler, Phys. Rev. Lett. 79, 22, 4445 (1997)

Mater. Res. Soc. Symp. Proc. Vol. 1613 © 2014 Materials Research Society
DOI: 10.1557/opl.2014.170

Impact of Natural and Synthetic Nanofibres Presence in Polymeric Composites on Mechanical Properties

Martin Seidl, Jiří Bobek, Jiří Habr, Petr Lenfeld, Luboš Běhálek
Technical University of Liberec, Czech Republic

ABSTRACT

This study deals with evaluation of mechanical properties (tensile, flexural and impact strength), that verified reinforcing potential of nanofibres in polymeric composites and their applicability in dependence on the filler content of nanofibres of natural and synthetic origin. Considering the hydrophilic nature of natural nanofibres and hydrophobic polypropylene matrix was applied chemical modification that ensures effective interlock of nanofibres with the matrix, namely maleic anhydride grafted polypropylene and ionic modifier. Polymeric nanocomposites were prepared by twin screw extrusion that made possible mixing of all three components together. After cooling in water bath the composites were cut on pellets and dried for further injection moulding. The specimens were made of two- or three-phase component systems that included PP matrix, coupling agent in the volume of 4 wt.% and reinforcing nanofillers in the volume of 2, 4 and 6 wt.%. The presence of nanoparticles and coupling agents had not unambiguous impact on analysed mechanical properties. Coupling agent presence was dominant for impact strength, however, flexural and tensile properties were influenced mainly by the nanofiller type and origin. The dispersed phase and compatibilizer improved flexural modulus but tensile modulus of pure PP were not achieved by three-phase composite, not even with the highest analysed nanoparticle load. Cellulose fibres proved as more proper than the synthetic ones not only for their biodegradability and renewability but for their better toughness as well.

INTRODUCTION

Polymers are widely used materials with very good processability, low density and cost compared to ceramics and metals. A large range of applicable additives and fillers of organic or inorganic origin that can be incorporated into polymer matrix, provide various modifications and improvement of final performance and also affect the total price of the composite. Properties are synergistically derived from all phases that are involved in the multiphase system. [01] Aimed modification of mechanical, chemical and physical properties and processability of polymer matrix is a source of materials with potential for hi-tech applications, e.g. electronics, constructions, aerospace, automotive, sporting goods, medicine, biotechnology [02] and thermoplastic based nanocomposites are one of very fast growing areas of nano-engineered materials due to need of lower filler load comparing to fillers in micro and/or macro scale that have comparable properties.

Polymer composites are usually two- or three-component systems. The continual phase is called the matrix and is formed by polymeric material in which is dispersed second discontinuous phase created by reinforcing fillers. The third component is usually a compatibilizer or coupling agent providing homogenous interface between the dispersed and continual phases [03].

Matrix as the continual phase determines characteristic of the composite by its molecular weight, transition temperatures, crystallinity degree, level of crosslink density and the power of interactions with filler surface [04]. The main role of the matrix is to transfer stress onto the fibres and protect the fibres against environmental influences and damages. To reach the full mechanical potential of fibres it is necessary for matrix to have better deformation resistance than fillers have [05].

Reinforcing nanoparticles have at least one dimension in the nano range (1-100 nm). The nanofillers can have one dimension (e.g. plates, laminas and/or shells), two dimensions (e.g. nanotubes, nanofibres, whiskers or sepiolite), or three dimensions (spherical particles) in the nanometer range. [06] Nanoparticle characteristics that primarily control the surface behaviour include the chemical composition and the resultant solubility, surface charge, size, shape, surface curvature, crystallinity, porosity, surface heterogeneity, roughness etc. [07] Particles in nanoscale have a different behaviour compared to bulk material. The first dissimilarity is caused by surface effects what means smooth properties scaling due to the fractions of atoms at the surface. The second one is a quantum effect that shows discontinuous behaviour due to quantum confinement effects in materials with delocalized electrons [08].

In general, nanoparticles provide reinforcing efficiency because of their high aspect ratios and surface area (high particle numbers per unit mass). As the size of a particle is reduced, the number of defects per particle is also reduced and mechanical properties rise proportionately. [09] Purpose of nanoparticle use is an effective reinforcement without loss of ductility, improve of impact strength and heat deflection temperature, enhancement in gas barrier properties with low filler content (< 5 wt.%) without changing of resin transparency and specific weight of final composite. Large improvements in mechanical properties can be obtained at much lower reinforcement levels comparing to particles in micro- or macroscale [10].

Nanoparticles do not disrupt the morphology of polymer matrix due to their nanoscale dimensions and low concentrations, however they affect molecular mobility (relaxation behaviour and thermal transitions temperatures) [11]. The forces between nanoparticles are classified into long range and short range forces. The basic forces are derived from usually attractive van der Waals forces and repulsive or attractive electrostatic interactions, whereas, the latter arise from steric, depletion and hydration interactions. These forces also cause the tendency of nanoparticles to agglomerate and then this cluster behave as a larger particle depending on its total size.The main challenge is related to homogeneous dispersion of nanofillers within a polymeric matrix. Uniform distribution of the nanofillers in the polymer matrix is one of the most important parameters that decide the properties of a nanocomposite. Poorly dispersed nanomaterial may degrade the final mechanical properties Better distribution and dispergation were reached when the nanocomposites were prepared by two screw extruder or kneader [07].

The *interfacial area (interphase)* is the most important zone for mechanical properties of the composite because controls the load transfer between matrix and reinforcing nanoparticles. In order to enhance polymer-to-nanofiller interactions, the chemical or physical modification of the nanoparticle surface is a widespread used strategy, especially with natural nanofibres and polysaccharidic nanoparticles etc. During the melt-processing, a covalent chemical bond can be formed at the polymer/nanofiller interface by grafting, reactive extrusion or cross-linking techniques. The best interactions are reached when using nanoparticles reaching the atomic or molecular dimensions [01].

Negative limitation that causes reducing of adhesion between polymer matrix and reinforced nanoparticles is the presence of *moisture* which could be caused by water molecules interfering with the hydrogen bonding interactions between the nanofibres and the matrix. Water also acts as a plasticizer [12].

Coupling agents are based on the concept that when two materials are incompatible, a third material with intermediate properties can bring the compatibility to the mixture. The coupling agents should enhance very low affinity between non-polar matrix and polar fillers and have two functions: to react with OH groups of the cellulose and to react with the functional groups of the matrix with the goal of facilitating stress transfer between the fibres and the matrix. The most widely used coupling agents include organosilanes, tryazine and maleic-anhydride. Adding the coupling agents to the matrix changes also the degree of filler dispersion [06].

Natural nanofibres are mainly composed of cellulose, whose elementary unit (anhydro d-glucose) contains three hydroxyl (OH) groups. These hydroxyl groups form intra and intermelecular bonds, causing all vegetable fibres to be hydrophilic. The properties of cellulose fibres are affected by many factors such as variety, climate, harvest, maturity, retting degree, decortications, disintegration, age of plant etc. Cellulose fibres are being used as potential reinforcing materials because of so many advantages such as availability and price independent on the price of rope, low weight, biodegradability, renewability, low abrasive nature, interesting specific properties, since these are waste biomass and exhibit good mechanical properties. Cellulose fibres also have some disadvantages such as moisture absorption, quality variations, low thermal stability and poor compatibility with the hydrophobic polymer matrix that are very often physically or chemically treated [13].

EXPERIMENT

Composite materials were prepared by cold granulation and the granulation line (Zemak, Poland) was equipped by two-screw extruder. As a matrix was used polypropylene obtained from Sumika Polymers Compounds (product name Thermofil PP E020M, without talc) with high melt flow index that is given for injection moulding processing. For influence study of different types of coupling agents the Fusabond P was chosen (DuPont® company) as a representative of maleic-anhydride group. The second coupling agent was Struktol® TPW 243 (Struktol Company of America) that denotes second generation compatibilizer that works by reactive extrusion which has the ability to bond organic and inorganic materials by a free radical source through an ionic mechanism initiated by hydrolysis. Both coupling agents were added in the volume of 4 wt. %.

Mechanical properties of nanocomposites were evaluated on specimens manufactured by injection moulding (IMM Arburg 270 S 400 - 100). Regarding to assumption wide utilization of nanocomposites were evaluated mainly strength parameters according to standards ISO 527 (ISO 527/1A/1), ISO 178 a ISO 179-1 in standard environment 23/50 (ISO 291). The mechanical characteristics as tensile strain at yield (σ_y, ISO 527/1A/50) and tensile elongation at break (ε_B, ISO 527/1A/50) were defined according to standard ISO 1873-2 using testing velocity of 50 mm/min, secant tensile modulus (E, (ISO 527/1A/1) was defined using testing velocity of 1 mm/min, flexural strength (σ_{fM}, ISO 178) and flexural modulus (E_f, ISO 178) measured using testing velocity of 2 mm/min and Charpy notched impact strength (a_{cA}, ISO 179-1/1eA) using hammer velocity of 2,9 m/s.

RESULTS AND DISCUSSION

All prepared samples followed the changes and impact of presence two types of coupling agents and changing nanofiller content in volume of 2, 4 and 6 wt. %. Firstly were performed tensile tests and the results are compared in figures 1, 2 and in table 1. In figure 1 can be observed the influence of two- or three-phases on the tensile modulus of multiphase composite system. The best results were reached with samples made of Sumika (pure PP). Behaviour of tensile modulus was similar when adding only 4 wt. % of both coupling agent types. The measured values decreased for about 13% compared to pure PP. The tensile modulus generally increases with increasing filler content but we did not reached value of pure PP matrix. Much better interactions were observed between PP matrix and nanoparticles of cellulose origin with Fusabond, however, the gained results are very close to each other when comparing the impact of coupling agents only. In figure 2 can be observed the course of flexural modulus. On the contrary to tensile modulus the flexural modulus of multiphase composite system was higher for all studied variation than the flexural modulus of pure PP. Comparing the impact of two-phase systems (PP + Struktol or Fusabond) better improvement was reached with Fusabond, however, the best results were gained when using cellulose nanoparticles with Struktol. The values were growing with increasing the filler content up to 4 wt. %. With higher nanoparticles load the values of flexural modulus were decreasing. The table 1 summarizes tensile, flexural and impact strain at yield, tensile elongation at beak, flexural strength and impact strength of V-notched specimens at temperature of +23°C and -35°C. From the strength viewpoint were the best results gained when measuring composites with cellulose nanoparticles Arbocel without bigger impact of coupling agent type. All values except impact strength gained by two- or three -phase composite systems were equal or better comparing to values of pure PP. Elongation at break decreased with increasing nanofiller content. The synthetic nanoparticles reached lower ductility than the natural one. From the impact strength viewpoint pure PP reached the best impact strength at both analysed temperatures. With increasing nanofiller load the toughness was decreasing. Influence of nanofiller type was not definite and deciding phase was the coupling agent. Interactions of nanoparticles with PP matrix improved by Struktol coupling agent proved as more impact resistant.

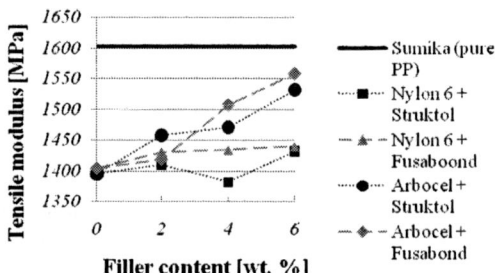

Figure 1. Tensile modulus of analyzed nanocomposites

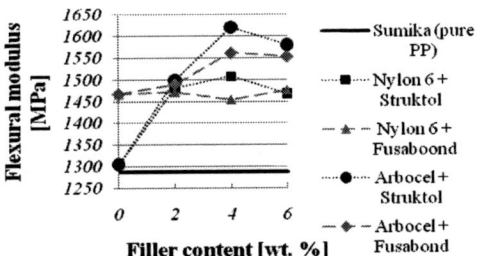

Figure 2. Flexural modulus of analyzed nanocomposites mechanical properties namely tensile

Table I. Mechanical properties of analyzed nanocomposites

Nanoparticle type	Coupling agent	Filler content	Tensile strain at yield [MPa]	Tensile elongation at break [%]	Flexural strength [MPa]	Impact strength, V-notched, 23°C [kJ/m2]	Impact strength, V-notched, -35°C [kJ/m2]
	Pure PP		27	39,4	35	8,8	3,8
Nylon 6	Struktol (4 wt.%)	0 wt.%	28	30,5	36	7,7	3,3
		2 wt.%	27	8,5	40	6,0	2,2
		4 wt.%	27	8,5	40	5,4	2,1
		6 wt.%	27	7,7	40	4,7	2,0
	Fusaboond (4 wt%)	0 wt.%	29	53,7	40	6,8	2,5
		2 wt.%	27	9,3	40	5,5	2,2
		4 wt.%	27	7,1	39	5,3	1,8
		6 wt.%	27	7,9	41	4,0	1,8
Arbocel	Struktol (4 wt.%)	0 wt.%	28	30,5	36	7,7	3,3
		2 wt.%	28	35,8	40	5,9	2,4
		4 wt.%	29	30,8	41	5,4	2,2
		6 wt.%	29	23,2	41	4,9	2,2
	Fusabond (4 wt%)	0 wt.%	29	53,7	40	6,8	2,5
		2 wt.%	28	30,2	40	5,5	2,3
		4 wt.%	28	25,0	41	4,8	2,0
		6 wt.%	29	15,8	41	4,3	2,1

CONCLUSION

This study devotes to evaluation of mechanical properties of composites reinforced with nanoparticles of synthetic and natural origin. Polypropylene was chosen as a matrix and natural cellulose and nanofibres made of Nylon 6 were chosen as reinforcement. Maleic anhydride grafted polypropylene (Fusabond) and ionic modifier (Struktol) were used as third phase that would ensure effective interlock of hydrophilic nanofibres with hydrophobic matrix. A very good dispersion was achieved by co-rotating twin-screw extruder that provided incorporation of nanoparticles into polymer matrix and blending the base mixture with the coupling agent.
The trend of increasing mechanical properties with increasing nanofiller concentration (up to 6 wt. %) was not unambiguously confirmed. The best results of tensile strain at yield were reached when using Arbocel nanoparticles without significant influence of different coupling agent type and similar results were gained for flexural strength. The impact strength of notched specimens

was influenced mainly by the coupling agent type at both tested temperatures (23°C and -35°C). More effective interphase was provided by Struktol and as tensile elongation at break the values of notched impact strength decreased with increasing nanofiller load. The natural nanofillers prove as more toughness compared to synthetic ones. Flexural modulus of polymer matrix was improved by both nanoparticle phase and interphase up to 4 wt.% of nanofillers. Higher nanofiller load caused the decrease of flexural modulus. These parameters were affected by adding the nanofillers and coupling agents in expected way. On the contrary the tensile modulus of three phase system did not reach the value of pure PP matrix in analyzed range even though that higher nanofiller load improve the tensile modulus of composite material. The most proper combination seems to be cellulose nanofillers with Struktol from the viewpoint of mechanical properties, however, the differences between values affected by both types of coupling agents are nearly negligible To improve the interactions between polymer matrix and reinforcing nanoparticles will be analyzed another type of coupling agent (e.g. based on silanes) or a chemical or physical surface treatment. The volume of used coupling agent must be considered as well because with increasing surface area of nanoparticles increasing volume of chemical compatibilizers must be added to reach high enough density of interphase.

AKNOWLEDGMENTS

This research was financially supported by Technology agency of the Czech Republic, concretely research project TA01010946 and by the Project OP VaVpI Centre for Nanomaterials, Advanced Technologies and Innovation CZ.1.05/2.1.00/01.0005. Required technical equipment was provided by Technical University of Liberec.

REFERENCES

1. E. Ritzhaupt-Kleissl, J. Haußelt and T. Hanemann. "Thermo-mechanical properties of thermoplastic polymer-nanofiller composites," *4M Network of Excellence,* (2008).
2. F. Hussain, M. Hojjati, M. Okamoto and R.E. Gorga. "Review article: Polymer-matrix nanocomposites, processing, manufacturing, and application: An overview," *Journal of Composites Materials* (2006).
3. N, Tucker and K. A. Lindsey. *An introduction to automotive composites.* Shrewsbury, U.K., Rapra Technology Ltd., viii, 194, (2002).
4. R. N. Rothon, *Particulate fillers for polymers.* Shawbury, U.K.: Rapra Technology Ltd., pp. 154-158, v. 12, no. 9. ISBN 185957310X, (2002).
5. S. Taj, A. Munawar and S. Khan, "Natural fiber-reinforced polymer composites," *Proc. Pakistan Acad. Sci.,* pp. 129, (2007).
6. J. Cuppoletti,"Nanocomposites and polymer with analytical methods,"in *Intech ,* pp. 3, (2001).
7. F. Somasundaran and S. Ponnurangam, *KONA Powder and Particle Journal,* 28, pp. 38, (2010).
8. C. Buzea, I. I. Pacheco and K. Robbie, 2, 4. (2007).
9. T. Theivasanthi and M. Alagar, *Nano Biomedicine Engineering,* (2011).
10. W.E. Gacitua, A. A. Ballerini and J. Zhang, "Polymer nanocomposites: Synthetic and matural fillers: A review, " *Maderas: ciencia y tecnología* 7 (3),159 (2005).
11. G. Schmidt and M. M. Malwitz, *Current Opinion in Colloid & Interface Science,* 8, 1, 103, (2003).

12. D. Vllasveld, "Fibre reinforced polymer nanocomposites", Ph.D. Eng thesis, Dutch Polymer Inst., Haveka, Delft University of technology, ISBN 978-90-9019883-5, 2005.
13. S. Kalia, A. Dufresne, B. M. Cherian, B. S. Kaith, L. Avérous, J. Njuguna a E. Nassiopoulos, *International Journal of Polymer Science,* 2011, 1,(2011).

Rheology in Polymers

Mater. Res. Soc. Symp. Proc. Vol. 1613 © 2014 Materials Research Society
DOI: 10.1557/opl.2014.171

Rheological Properties of Multi-Block Associative Polyelectrolytes Obtained by Nitroxide-Mediated Solution Polymerization

Alejandro Coronado,[1] Areli I. Velazquez[2] and Enrique J. Jiménez[2]

[1] Facultad de Ciencias Químicas, Universidad Autónoma de Coahuila, Venustiano Carranza Blvd and Ing. José Cárdenas Valdez St., 25280, Saltillo, Coahuila, México
[2] Centro de Investigación en Química Aplicada (CIQA), 140 Enrique Reyna Blvd, 25294, Saltillo, Coahuila, México

ABSTRACT

A multi-block associative polyelectrolyte based on poly(methacrylic acid-ra-styrene) [MAA-S] and poly(octadecyl methacrylate) [ODMA] was synthesized through stepwise nitroxide-mediated solution polymerizations. The obtained polymer has a heptablock copolymer structure, alternating MAA-S as hydrophilic blocks (theoretical degree of polymerization [DP^T] of 250), and ODMA as hydrophobic blocks ($DP^T = 15$). Rheological properties, in the linear-response regime, of aqueous solutions (polymer content = 1.5 wt.%) were studied as a function of the amount of blocks on the polymer using steady-shear and creep-compliance experiments. Rheological experiments demonstrate that the viscoelastic behavior of the polymer bearing an ODMA block in terminal position greatly differs from that of the polymer with MAA-S block terminations. The former behaves as a newtonian fluid on a wider range of shear rates than the latter, which exhibit a shear-thinning behavior, even at low shear rates, independently of the molecular weight and number of blocks.

INTRODUCTION

The tunable viscoelastic properties and the high thickening effect, stronger and at lower concentrations than that achievable with neutral polymers, exhibited by associative polyelectrolytes, like the hydrophobically modified alkali-soluble emulsions polymers [H.A.S.E.], on aqueous solutions make them appropriate as rheology modifiers in various aqueous formulations for industrial applications [1–4]. This functionality combined with the precise control of the polymer architecture offered by the controlled/living radical polymerization techniques [5], of which worth mentioning are the Atom Transfer Radical Polymerization [ATRP] [6], Reversible Addition-Fragmentation chain-Transfer [RAFT] [7] and Nitroxide-Mediated Polymerization [NMP] [8–11], has allowed us to study and understand the influence of the molecular architectural characteristics (i.e., molecular weight, molecular molar mass distribution (dispersity), functionality, composition, block topology, hydrophobic/hydrophilic ratio) on the viscoelastic behavior of these polymers on aqueous solution. Herein, the NMP technique stands out because can be carried out under relatively undemanding conditions (such as reagent purity), the use of pure organic systems, no use of transition metals-based catalysts and the low potential for odor and discoloration [5].

In this study the ability of the NMP technique to chain extend a polymer is exploited in order to obtain a multi-block associative polyelectrolyte [12,13] based on monomers used on H.A.S.E. polymers, methacrylic acid [MAA] and styrene [S] forming the hydrophilic blocks

[14], and octadecyl methacrylate [ODMA] as the hydrophobe blocks. This technique is combined with the advantages of the solution polymerization technique [15] in the control of the temperature, thermal dissipation and viscosity of the reaction mixture. Rheological properties were studied as a function of the amount of blocks on the polymer using steady-shear and complementary creep-compliance experiments.

EXPERIMENTAL DETAILS

Materials

Analytical grade reagents were used, without further purification, as supplied by each company. Ammonium hydroxide [NH_4OH], acetone-d_6, MAA, Styrene and ODMA were supplied by Aldrich. 4-hydroxy-2,2,6,6-tetramethylpiperidinyl-1-oxy [OH-TEMPO] was from Ciba Specialty Chemicals Corporation. Dimethylformamide [DMF] and benzoyl peroxide [BPO] were from Sigma-Aldrich. Tetrahydrofuran [THF] was from Tedia. Ethyl ether and acetone were from Fermont. UHP-grade nitrogene was from Infra. Distilled deionized water was obtained from a Cole-Parmer ion-exchange system.

Synthesis

The copolymer under investigation was synthesized by stepwise Nitroxide-Mediated Solution Polymerization [16]. The first block synthesis, aiming for a DP^T close to 250 ($M_N \approx$ 25000 g mol^{-1}), was performed as follows: a homogeneous mixture of 5.474 g (5.25 x 10^{-2} mol, initial molar ratio, $f_{S,0}$ = 0.5) of S, 4.526 g (5.25 x 10^{-2} mol) of MAA, 6.88 x 10^{-2} g of OH-TEMPO (4 x 10^{-4} mol), as control agent, and 7.45 x 10^{-2} g of BPO (3.07 x 10^{-4} mol, nitroxide:initiator ratio of 1.3), as initiator, in a solvent (DMF, 10 g, solution polymerization at 50 wt.%) was deoxygenated with a nitrogen stream for 15 min at room temperature. The so-formed solution was then introduced into a 25 mL three-necked jacketed reactor connected to a Julabo F25 heating circulator to control the temperature. The reaction system was equipped with a reflux condenser, thermocouple, magnetic stirring and a nitrogen inlet to maintain an inert atmosphere. All polymerization reactions were carried out at 135 °C (no heating ramp procedure) during 8 hr. Samples were periodically withdrawn in order to determine polymerization kinetics. Consecutive polymerizations were carried out with the same reaction technique previously described using the purified copolymer from the preceding step (macroinitiator) and adding it into the reactor with DMF solvent (solution polymerization at 20 wt.%) and defined amounts of monomer (ODMA or MAA and S) (see Table I) targeting a DP^T of 15 for hydrophobic ODMA blocks ($M_N \approx$ 5000 g mol^{-1}) or a DP^T close to 250 ($M_N \approx$ 25000 g mol^{-1}) for MAA-S hydrophilic blocks. Resulting polymerization batches were rinsed with distilled water three times to remove DMF. After drying, copolymers were purified from monomer dissolving them in acetone and precipitated in ethyl ether [14], and after filtration, they were dried under reduced pressure at 50 °C for 24 hr using a lab-line Duo-Vac vacuum oven manufactured by LabLine Corp and freeze-dried in a Freezone 6 Freeze-Dryer System of Labconco during 12 hr. The overall monomer conversion (x) was determined by gravimetric analysis of purified copolymers.

Characterization

Gel permeation chromatography (GPC) was used to determine the number average molar mass of polymers [MN] and the molar mass distribution or dispersity [Đ] of the polymers. Samples were prepared dissolving 1 mg of the polymer in 1 mL of HPLC grade tetrahydrofuran (THF) without stabilizer. GPC analysis was performed on an Agilent 1100 series HPLC system coupled to GPC-SEC columns. Polystyrene was used as calibration standards for GPC. THF was used as an eluent and the flow rate was set to 1.0 mL min^{-1}. ^{1}H-NMR spectra of all the polymers were recorded on a JEOL Eclipse NMR spectrometer operating at 300 MHz at room temperature, using acetone-d_6 as solvent. The molar ratio of the styrene in the copolymers was determined by integrating the signal peaks corresponding to the vinylic protons, using the broad peak between 6.5 and 7.5 ppm as an internal reference (five aromatic H for S and one vinylic H for the S monomer).

Rheological Measurements

Aqueous solutions of 1.5 wt.% polymer were prepared by dissolution of a known amount of polymer in deionized distilled water and solubilized to a pH close to 9 using NH$_4$OH. After 24 hr of magnetic stirring for homogenization, the solutions were subjected to sonication in a Branson 2510 Ultrasonic Cleaner in order to eliminate the air bubbles trapped in the viscous fluids, followed by a 24 hr rest period. Rheological studies were performed on an Anton Paar Physica MCR-501 controlled stress rheometer coupled to a Julabo F-25 heating bath at 25±0.04 °C. The cone and plate fixture (diameter = 50 mm, angle = 2°) was used. Steady-shear viscosity measurements were carried out in order to characterize the viscoelastic behavior of aqueous solutions at shear rates from 0.01 to 1000 s^{-1}. Creep-compliance experiments were additionally performed on samples that were viscous enough to provide a meaningful analysis using small stresses so that the sample was within the linear viscoelastic region.

DISCUSSION

Table I summarizes the characterizations of the polymer synthesis. As proven by Figure 1a, the polymerization follows the expected kinetic behavior of the controlled/living radical polymerization techniques: the number average molar mass, M$_N$, increases linearly with conversion. With this, it can be asserted that the M$_N$ of the following polymers are within their expected values according to the obtained conversions (see Figure 1b). It is also seen that monomer conversions lower as polymerizations advance, owe to intrinsic limitations of the NMP technique (i.e., termination of polymerization by combination of polymer radicals, chain-chain terminations or chain-nitroxide terminations [13]). Styrene compositions in the final polymers were obtained from ^{1}H-NMR spectra.

Table I. Characteristics of the synthesized polymers.

Polymer	Feed Composition (g)				Expected M$_N$ at 100% conversion (g mol^{-1})	Experimental M$_N$ (g mol^{-1}) [b]	Conversion [x] (%) [c]	S Content (mol %) [d]
	MI [a]	MAA	S	ODMA				
MAA-S	-	4.562	5.474	-	25000	15817	61	40
Diblock	2.85	-	-	0.9009	20817	18901	57	36
Triblock	3.1	1.855	2.244	-	43901	33652	54	39
Tetrablock	5.11	-	-	0.7598	38652	36599	47	37

Pentablock	4.5	1.391	1.682	-	61599	49520	39	39
Hexablock	4.2	-	-	0.4263	54520	51353	31	37
Heptablock	3.3	0.727	0.879	-	76353	56980	12	37

[a]Macroinitiator; [b]obtained by GPC; [c]determined by gravimetric analysis; [d]calculated from ^{1}H-NMR spectra

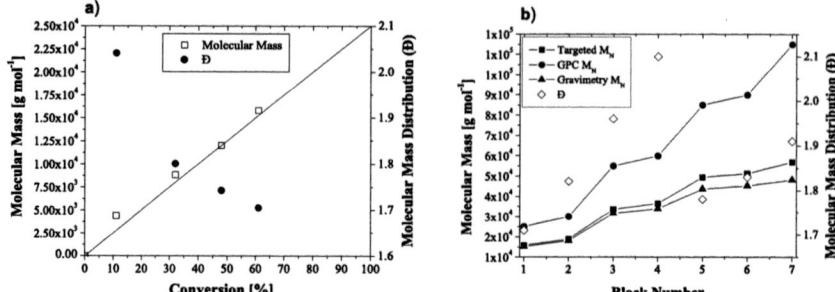

Figure 1. a) - M_N versus monomer conversion, x, for NMP of S and MAA. The squares and the solid line represent the linear behavior of the controlled polymerizations. **b)** - M_N and molar mass distribution, Đ, for each block polymer.

Rheological Studies

The polymers were found to form a physical hydrogel via the intermolecular hydrophobic association of the ODMA blocks and coil expansion phenomena caused by the ionic charge of MAA groups. Analyzing Figure 2a, it can be inferred the influence of the hydrophobic position in the rheological behavior: steady-shear tests showed that the zero-shear viscosity, η_0, increased with the number of blocks, with a pronounced increment each time a new hydrophobic group was added. The zero-shear viscosity is determined by extrapolation of the apparent viscosity in the limit of zero shear rates.

Rheological results show the difference between the viscoelastic behavior of the polymer bearing an ODMA block in terminal position and that of the polymer with MAA-S block terminations. The former behaves as a newtonian fluid on a wider range of shear rates than the latter, which exhibit a shear-thinning behavior, even at low shear rates, independently of the molecular weight and number of blocks. As the MAA in the hydrophilic blocks of the polymers is ionized, the coulombic repulsive force due to negative charges along the chain causes it to adopt an extended conformation, consequently increasing its hydrodynamic volume in the solution. For the hydrophobic blocks, the expansion of the chain backbone at high pH consequently allows a significant crossover from the predominantly intramolecular associations of the hydrophobic groups from within the same polymer chain to more intermolecular associations of hydrophobes from neighboring chains by Van der Waals forces, forming a transient network structure that increases the solution viscosity [17,18]. The synergy between the coil expansion phenomenon and the strong hydrophobic junctions significantly enhances the thickening efficiency of the polymer, even with relatively low molecular weight and polymer concentration [19].

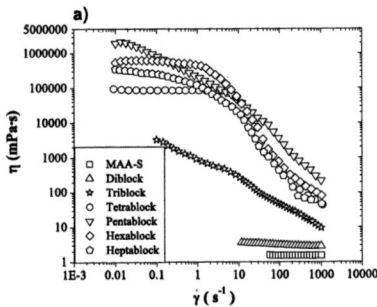
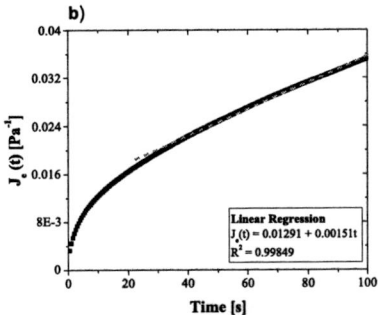

Figure 2. a) - Apparent viscosity, η, versus shear rate, γ, for all synthesized polymers. **b)** - Compliance, $J_e(t)$, versus time curve for 1.5 wt.% MAA-S-ODMA hexablock copolymer solution at pH 9. The broken line represents the linear fit at the linear part of the curve.

Figure 2b shows a creep-compliance test performed to the hexablock copolymer solution. The creep-compliance measurements are based on the sudden application of a constant stress to the fluid being tested and on the monitoring of the resulting deformation as a function of time [19]. In these experiments, a rapid change in the compliance is initially observed, followed by a smooth increase that becomes linear at longer times. The values of the zero-shear viscosity, η_0, the plateau module, G_0, and the terminal relaxation time, T_R can be obtained from the slope and the intersection of the linear part of the compliance (J_e) with the ordinate axis, according to the following equations [20,21]:

$$J_e(t) = \frac{1}{G_0} + \frac{t}{\eta_0} \tag{1}$$

$$T_R = \frac{\eta_0}{G_0} \tag{2}$$

For the sample shown in Figure 2b, η_0 was determined to be 662.251 Pa·s and G_0 is 77.45 Pa, therefore T_R was calculated to be 8.54 s. Rheological data of all polymers is shown in Table II. Both the steady-shear and the creep-compliance tests are in good agreement giving similar values for the zero-shear viscosity [20].

Table II. Rheological data for synthesized polymers.

Polymer	η_0 [Pa·s] (Steady-Shear/Creep-Compliance)	G_0 [Pa]	T_R [s]
MAA-S	1.56×10^{-3} / -	-	-
Diblock	3.43×10^{-3} / -	-	-
Triblock	3.07 / -	-	-
Tetrablock	91.56 / 111.3	21.91	4.17
Pentablock	1990 / 2057.37	171.73	12.05
Hexablock	602.37 / 662.251	77.45	8.54
Heptablock	330.785 / 270.81	53.43	6.19

Creep-compliance tests proved that the terminal relaxation time, T_R, reaches a maximum at the pentablock polymer, however in the following polymers, this time decreases. The terminal

147

relaxation time is often related to the time that two hydrophobic groups interact and held together while the plateau modulus, G_0, is related to the number of bonds between hydrophobic groups present in the solution at a given time. So, as expected, T_R and G_0 increase as increasing the number of blocks, however after the pentablock copolymer, both these values and the solution viscosity decrease. This can be explained by the fact that the amount of hydrophobic groups after the pentablock is such that it favors much more intramolecular associations shrinking the polyelectrolyte into a compact coil, hence the viscosity drop.

CONCLUSIONS

In this research it is reported for the first time the synthesis of a heptablock associative polyelectrolyte using the stepwise nitroxide-mediated solution polymerization. The rheological behavior was studied for each multiblock copolymer and it was observed the strong influence of the topology (molecular mass and position of hydrophobic groups) in the viscoelastic response of these polymers in aqueous solution. Rheological experiments have shown that the viscoelastic behavior of the polymer bearing an ODMA block in terminal position is different from that of the polymer with MAA-S block terminations. Since the hydrophilic segments of the copolymer exhibited a shear-thinning behavior and the hydrophobic groups displayed a newtonian behavior, this system presents a rich behavior regarding its rheological properties in aqueous solutions as a function of the position of the hydrophobic groups. The pentablock copolymer exhibited the higher thickening efficient of all polymers due to the optimal amount of strong hydrophobic junctions, and the synergy with the coil expansion phenomenon, typical of associative polyelectrolytes.

ACKNOWLEDGMENTS

Research financially supported by the Consejo Nacional de Ciencia y Tecnología of México (CONACyT) through project number 105712. One of the authors (A.C.) thanks C. Rivera Vallejo and H. Maldonado Textle for their valuable help in the experimental part of this research.

REFERENCES

1. C. Tsitsilianis, T. Aubry, I. Iliopoulos and S. Norvez, *Macromolecules* **43**, 7779 (2010).
2. A.C. Lara Ceniceros, C.C. Rivera Vallejo and E.J. Jiménez Regalado, *Polym. Bull.* **59**, 499 (2007).
3. A. Hill, F. Candau and J. Selb, in *Trends in Colloid and Interface Science V*, edited by M. Corti and F. Mallamace (Dietrich Steinkopff Verlag GmbH & Co. KG, Strasbourg Cedex, France, 1991), p. 61–65.
4. E.J. Jiménez Regalado, G. Cadenas Pliego, M. Pérez Álvarez and Y. Hernández Valdez, *Macromol. Res.* **12**, 451 (2004).
5. *Controlled Radical Polymerization Guide*: ATRP|RAFT|NMP, edited by Aldrich Materials Science (2011), p. 52.
6. K. Matyjaszewski and J. Xia, in *Handbook of Radical Polymerization*, edited by K. Matyjaszewski and T.P. Davis, (WILEY-INTERSCIENCE: John Wiley and Sons, Inc., 2002), p. 523.

7. J. Chiefari, Y.K.B. Chong, F. Ercole, J. Krstina, J. Jeffery, T.P.T. Le, R.T.A. Mayadunne, G.F. Meijs, C.L. Moad, G. Moad, E. Rizzardo and S.H. Thang, *Macromolecules* **31**, 5559 (1998).
8. E. Rizzardo and D.H. Solomon, *Aust. J. Chem.* **65**, 945 (2012).
9. G. Moad and D.H. Solomon, *The Chemistry of Radical Polymerization*, 2nd ed (Elsevier Ltd, 2006) p. 413.
10. M.K. Georges, R.P.N. Veregin, P.M. Kazmaier and G.K. Hamer, *Macromolecules* **26**, 2987 (1993).
11. C.J. Hawker, A.W. Bosman and E. Harth, *Chem. Rev.* **101**, 3661 (2001).
12. A.S. Kimerling, W.E. (Skip) Rochefort and S.R. Bhatia, *Ind. Eng. Chem. Res.* **45**, 6885 (2006).
13. R.B. Grubbs, *Polym. Rev.* **51**, 104 (2011).
14. C. Dire, B. Charleux, S. Magnet and L. Couvreur, *Macromolecules* 40, 1897 (2007).
15. G. Odian, *Principles of Polymerization*, 4th ed (WILEY-INTERSCIENCE: John Wiley and Sons, Inc., 2004) p. 297.
16. C. Lefay, B. Charleux, M. Save, C. Chassenieux, O. Guerret and S. Magnet, *Polymer* 47, 1935 (2006).
17. V. Tirtaatmadja, K.C. Tam and R.D. Jenkins, *Macromolecules* **30**, 3271 (1997).
18. A.A. Abdala, PhD Thesis, North Carolina State University, 2002.
19. P. Kujawa, A. Audibert Hayet, J. Selb and F. Candau, *J. Polym. Sci. Part B Polym. Phys.* **42**, 1640 (2004).
20. P. Kujawa, A. Audibert Hayet, J. Selb and F. Candau, *Macromolecules* **39**, 384 (2006).
21. M.T. Popescu, C. Tsitsilianis, C.M. Papadakis, J. Adelsberger, S. Balog, P. Busch, N.A. Hadjiantoniou and C.S. Patrickios, *Macromolecules* **45**, 3523 (2012).

Mater. Res. Soc. Symp. Proc. Vol. 1613 © 2014 Materials Research Society
DOI: 10.1557/opl.2014.173

Rheological Properties of Associative Polyelectrolytes Synthesized by Solution Polymerization

Areli I. Velazquez[1], Alejandro Coronado[2] and Enrique J. Jiménez[1]
[1]Centro de investigación en Química Aplicada CIQA, Blvd. Enrique Reyna 140, 25294, Saltillo, México.
[2]Facultad de Ciencias Químicas, Universidad Autónoma de Coahuila, Blvd. Venustiano Carranza and C. Ing. José Cárdenas Valdez, 25280, Saltillo, México

ABSTRACT

Water-soluble associative polyelectrolytes of methacrylic acid [MAA] and ethyl acrylate [EA] (1:1 molar ratio), hydrophobically modified with small amounts of stearyl metacrylate [MM_{18}], were synthesized by means of solution polymerization. Polyelectrolytes with two different molecular structures: multisticker, with hydrophobic groups randomly distributed along the hydrophilic chain and combined, with the hydrophobic groups along the chain and as terminal groups of the backbone, were obtained. Steady shear behavior and linear viscoelastic properties were studied as a function of polymer microstructure and hydrophobic group concentrations on salt-free aqueous solution using a cone-and-plate rheometer. Concentration regimes were obtained for each synthetized polymer. Viscoelastic study shows that the maximum thickening effect corresponds to the combined structure followed by multisticker structure. These polyelectrolytes exhibit high thickening power on aqueous solutions due to the synergy between the hydrophobic attractive interactions and coil expansion phenomena.

INTRODUCTION

Water soluble polymers are of great importance at industrial level (as thickening and gelling agents in paints, cosmetics, oil fluids, etc.). Their thickening power of these polymers are directly related to the hydrodynamic volume of the macromolecule in solution, therefore high molecular weights are commonly observed. However, their main drawback is suffering mechanical degradation at high shear rates, which leads to a irreversible decrease in their viscosity.

4 decades ago appeared a new class of water-soluble polymers: associative polymers, which have good thickening properties even with relatively small molecular weights [1]. Water-soluble associative polymers are macromolecules composed of a hydrophilic monomer (in excess) and a hydrophobic monomer (small proportion, 0.5-5 mol%) distributed along the chain or as terminal groups. In aqueous solution and above a certain polymer concentration, the hydrophobic groups interact and form tridimensional transient networks of intra and intermolecular associations that increases the viscosity of the solution

Associative polymers are classified according to the hydrophobe position with respect to the hydrophilic backbone as: multisticker, microstructure with hydrophobic blocks randomly distributed along the hydrophilic chain [1-3]; telechelic, in which the hydrophobic groups are located in terminal position; and combined, displaying hydrophobes at both inside and terminal positions of the backbone [4-8].

Another way to increase their viscosity in aqueous solution is adding ionic charges in the polymer, i.e. polyelectrolytes. In this kind of associative polymer, ionic charges cause a repulsive force along the chain, which adopts a larger hydrodynamic volume, hence increasing the viscosity o the solution.

HASE (Hydrophobically Alcali Soluble Emulsion) polymers are one of the most widely used rheology modifiers in the paintings and coatings industries. HASE polymers presents a multisticker-type structure and are synthesized via emulsion polymerization, comprising of methacrylic acid (MAA), ethyl acrylate (EA) and of 1-3 mol% of hydrophobic macromonomer randomly distributed along the chain. At acidic pH (2-5), these polymers are insoluble in water, neutralizing them with a base (pH~ 9) MAA acid groups are ionized and polymer becomes soluble in water, at this moment the polymer chain behaves as an anionic polyelectrolyte which expands and, together with the hydrophobic groups, form a network structure that significantly increases the viscosity of the solution, its thickening power is present even at low concentrations of polymer (~ 2% weight). By the required type of synthesis telechelic and combined type structures can not be obtained, so we propose an alternative method of synthesis, solution polymerization, which also allows us to increase the solids content, yielding until 70% of solids.

EXPERIMENTAL DETAILS

Materials

All reagents were analytical grade and were used, without further purification, as supplied by each company. Methacrylic acid (MMA), ethyl acrylate (EA), stearyl acrylate (MM_{18}), 1-octadecanol and ammonium hydroxide were from Sigma-Aldrich. 4,4′-azobis(4-cyanovaleric acid) (ACVA), N,N′-dicyclohexylcabodiimide and 4-dimethylaminopyridine were from Fluka. Tetrahydrofuran (THF) was from Tedia. Methanol and acetone were from J.T. Baker and hexane was from Jalmex. Distilled and deionized water was obtained using an ionic exchange columns system of Cole-Parmer.

Hydrophobic Initiator Synthesis

The used method consists to functionalize the commercial azo-initiator 4,4-azobis-cyanovaleric acid (ACVA) with an esterification reaction between the ACVA and octadecanol as reported by Perez et al. [4].

Polyelectrolytes Synthesis

All polyelectrolytes, including references without hydrophobes groups, were synthesized via solution polymerization, using ethanol as dissolvent and setting a weight concentration at 70%. MAA and AE were used as the initial hydrophilic monomers and stearyl methacrylate as the hydrophobic monomers. 0.32 mol% of ACVA or previously synthesized hydrophobically modified $ACVA_{18}$ initiators was also used. Once polymerization is complete, the obtained polymer is dissolved in THF and then is precipitated in hexane. Finally, the polymer is dried under vacuum at 50 °C during 24 hours. Table I summarizes the nomenclature and characteristics of synthesized polymers.

Table I. Synthesized Polymers

Sample	Polyelectrolyte	Hydrophobe length (# of carbons)	Hydrophobe monomer concentration
M-18 (1)	Multisticker	18	1 mol%
M-18 (2)	Multisticker	18	2 mol%
C-18 (1)	Combined	18	1 mol%
C-18 (2)	Combined	18	2 mol%

Sample preparation

Aqueous solutions were prepared by weighing defined quantities of polymer and dissolving them in deionized water. In order to make solubilization possible, the samples were neutralized at a pH of approximately 9 adding dropwise an ammonium hydroxide solution (28-30% NH_3 content). Samples were homogenized under stirring for 24 hours, followed by a 24-hours rest period prior to each measurement.

Rheological measurements

The rheological studies were performed on an Anton Paar Physical MCR-501 rheometer, using the cone-plate geometry (50 mm 2°). Steady shear tests were carried out at 25±0.04 °C setting an interval of shear rate measurements from 0.01 to 1000 s-1. The zero-shear viscosity (η_0) was obtained by extrapolation of the apparent viscosity at very low shear rates.

DISCUSSION

As previously indicated, thickening effect of associative polymers are shown above a certain concentration of polymer in the solution, above this concentration hydrophobic intermolecular interactions occurs.

It is therefore of extreme importance, a study that shows the different concentration regimes for each polyelectrolyte; with the study can be determined the optimal concentration required to achieve a good thickening behavior of these polyelectrolytes.

It is known that if the concentration of polymer C is lower than the critical aggregation concentration (C*) then hydrophobic interactions only occur within the same molecule, favoring intramolecular interactions, but if the polymer concentration is increased until C>C*, more intermolecular interactions are favored, improving the thickening power [9]. In the case of associative polymers, the critical aggregation concentration is represented as C_η (this value is determined when the viscosity of the associative polymer exceeds that of the control reference polymer), the concentration of crosslinking with C_T and concentrated regime range with C_C.

Before analyzing the results further, it is useful to recall the main features of the rheological behavior of unmodified polymers. When discussing the dynamic properties of solutions of linear polymer chains, four concentrations regimens are generally distinguished [10-12]:

(i)   The dilute regime C<C* for which the zero-shear viscosity is of the order of that of the solvent.

(ii)   The semidilute unentangled regime C*<C< Ce. In this regime, viscoelasticity of the solution is controlled by the Rouse dynamic and the viscosity increases moderately. Note that the ratio of Ce/C* is generally of the order of 5 – 10. [25, 26].
(iii)   The semidilute entangled regime Ce<C<C**. The reptation model describes the viscosity properties in this regime and the viscosity follows a power law of the polymer concentration with an exponent close to 4. The plateau modulus is proportional to the density of entanglements.
(iv)   The concentrated regime stars beyond C**; the reptation model still holds but the scaling behavior becomes different.

Figure 2 shows the log-log variation of the steady-state viscosity (η) as a function of shear rate ($\dot{\gamma}$) for the combined type polyelectrolyte C-18 (1) at different concentrations in aqueous solution (C). In the dilute regime (C $\leq$ 0.3 wt%), the system behaves as newtonian, that is, there is no detectable variation of η with $\dot{\gamma}$, increasing further the concentration (C $\geq$ 0.7 wt%), the systems viscosity exhibits a shear thickening behavior and then a shear thinning behavior at higher shear rates. The shear thinning behavior increases as the polymer concentration does. The same behavior is observed in the other studied polyelectrolytes.

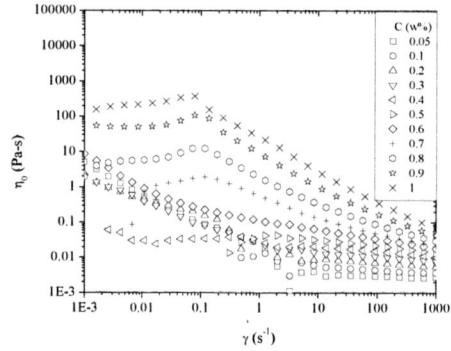

Figure 1. Steady-state viscosity η versus $\dot{\gamma}$ for C-18(1) sample at various concentrations in water (C).

In order to obtain the concentration regimes of these polyelectrolytes, the rheological tests was performed conducting a sweep of polymer concentrations from 0.05 to 14 wt% for the reference polyelectrolyte and concentrations from 0.005 to 1 wt% for combined and multisticker type polyelectrolytes, concentration which already exhibited high viscosities.

Figure 2 shows the log-log variation of the viscosity extrapolated to zero shear rates (η_0) as a function of the concentration of polymer (C) for multisticker type polyelectrolytes. For the reference polyelectrolyte, it is observed that the C* value is 0.3 wt%, while the value of C_e is approximately 5 wt%, it is also observed that above of C_e (semidilute entangled regime) the values follows a straight line with a slope of about 4, typical behavior of other associative polymers. For the M-18(1) and m-18(2) multisticker polyelectrolytes, its C_e value were found at

0.1 and 0.4 wt%, respectively, a lower concentration than the needed for the reference polyelectrolyte. it is also observed that above of C_e (semidilute entangled regime) the values follows a straight line with a slope of about 4, typical behavior of other associative polymers.

It is further noted that the amount of hydrophobic groups directly influences the viscosity of the solutions, showing differences of one order of magnitude approximately between similar samples with different concentrations of hydrophobic.

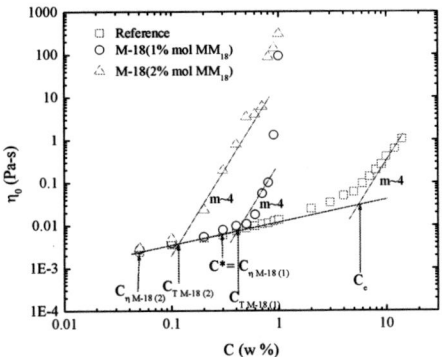

Figure 2. η_0 as a function of C for Reference, M-18 (1) and M-18 (2) polyelectrolytes.

Figure 3 shows the log-log variation of the viscosity extrapolated to zero shear rates (η_0) as a function of the concentration of polymer (C) for combined type polyelectrolytes. As before, it is observed the influence of the amount of hydrophobic groups in the solution viscosity. Here, C_e value were found at 0.08 and 0.4 wt%, for C-18(1) and C-18(2), respectively.

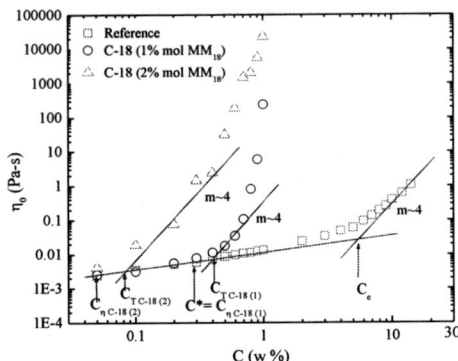

Figure 3. η_0 as a function of C for Reference, C-18 (1) and C-18 (2) polyelectrolytes.

155

CONCLUSIONS

Associative polyelectrolytes were synthesized with different amount of hydrophobic groups. There were obtained multisticker and combined-type polyelectrolytes. With the steady-shear rheological study it could be obtained their concentration regimes and the influence of the microstructure and the amount of hydrophobic groups.

The amount of hydrophobic groups per chain directly influences the concentrations at which regimes are presented and, therefore, the thickening power of the polyelectrolytes. It was also found that the microstructure of associative polyelectrolytes also influences these regimes, founding higher viscosities in the combined type microstructure than those of the multisticker type.

ACKNOWLEDGMENTS

The authors wish to thank the Consejo Nacional de Ciencia y Tecnología (CONACyT) for the financial support to this work, through project 105712.

REFERENCES

1. F. Candau and J. Selb, *Advances in colloid and interface science* 79, 149 (1999).
2. E. J. Jiménez, J. Selb and F. Candau, *Macromolecules* 33, 8720 (2000).
3. E. J. Jiménez, J. Selb and F. Candau, *Macromolecules* 32, 8580 (1999).
4. E. J. Jiménez, G. Cadenas, M. Pérez and Y. Hernández, *Polymer* 45, 1993 (2004).
5. E. J. Jiménez, G. Cadenas, M. Pérez and Y. Hernández, *Macromolecular Research* 12, 451 (2004).
6. A. C. Lara, C. Rivera and E. J. Jiménez, Polymer Bulletin 58, 425 (2007).
7. J. C. Rico and E. J. Jiménez, *Polymer Bulletin* 62, 57 (2009).
8. V. J. González and E. J. Jiménez, *Polymer Bulletin* 62, 727 (2009).
9. P. Williams, *Handbook of Industrial Water Soluble Polymers*. Primera Edición. UK, Blackwell Publishing Ltd. p.2 (2007).
10. P. G. Gennes, *Scaling Concepts in Polymer Physics*, Cornell University Press: London (1979).
11. W. W. Graessley, *Polymer* 21, 258 (1980).
12. R. H. Colby, M. Rubinstein and M. J. Daoud, *Phys.II France* 4, 1299 (1994).

AUTHOR INDEX

SUBJECT INDEX

CPSIA information can be obtained at www.ICGtesting.com
Printed in the USA
LVOW10*0229160414

381872LV00002B/3/P